Optical Techniques for the Determination of Nitrate in Environmental Waters: Guidelines for Instrument Selection, Operation, Deployment, Maintenance, Quality Assurance, and Data Reporting

By Brian A. Pellerin, Brian A. Bergamaschi, Bryan D. Downing, John Franco Saraceno, Jessica D. Garrett, and Lisa D. Olsen

Chapter 5 of
Section D, Water Quality
Book 1, Collection of Water Data by Direct Measurement

Techniques and Methods 1–D5

U.S. Department of the Interior
U.S. Geological Survey

U.S. Department of the Interior
SALLY JEWELL, Secretary

U.S. Geological Survey
Suzette M. Kimball, Acting Director

U.S. Geological Survey, Reston, Virginia: 2013

For more information on the USGS—the Federal source for science about the Earth, its natural and living resources, natural hazards, and the environment, visit http://www.usgs.gov or call 1–888–ASK–USGS.

For an overview of USGS information products, including maps, imagery, and publications, visit http://www.usgs.gov/pubprod

To order this and other USGS information products, visit http://store.usgs.gov

Suggested citation:
Pellerin, B.A., Bergamaschi, B.A., Downing, B.D., Saraceno, J.F., Garrett, J.A., and Olsen, L.D., 2013, Optical techniques for the determination of nitrate in environmental waters: Guidelines for instrument selection, operation, deployment, maintenance, quality assurance, and data reporting: U.S. Geological Survey Techniques and Methods 1–D5, 37 p.

Contents

Figures

Tables

Conversion Factors

SI to Inch/Pound

Multiply	By	To obtain
	Length	
centimeter (cm)	0.3937	inch (in.)
millimeter (mm)	0.03937	inch (in.)
nanometer (nm)	0.00000003937	inch (in.)
	Volume	
liter (L)	33.82	ounce, fluid (fl. oz)
	Mass	
milligram	0.00003527	ounce, avoirdupois (oz)

Temperature in degrees Celsius (°C) can be converted to degrees Fahrenheit (°F) as follows:

°F=(1.8×°C)+32

Abbreviations

ACS	American Chemical Society
ASTM	American Society for Testing and Materials
AU	absorbance units
DCP	data collection platform
DIW	deionized water
DOC	dissolved organic carbon
DOM	dissolved organic matter
FDOM	chromophoric dissolved organic matter fluorescence
FSP	Fundamental Science Practices
Hz	hertz
IBW	inorganic-grade blank water
IHSS	International Humic Substances Society
MSDS	material safety data sheet
N	nitrogen
NFM	National Field Manual
NFSS	National Field Supply Service
NWIS	National Water Information System
NWQL	National Water-Quality Lab
NTU	nephelometric units
OWQ	Office of Water Quality
PSU	practical salinity units
USGS	U.S. Geological Survey
UV	ultraviolet
VDC	volts direct current
VIS	visible
W	watts

Optical Techniques for the Determination of Nitrate in Environmental Waters: Guidelines for Instrument Selection, Operation, Deployment, Maintenance, Quality Assurance, and Data Reporting

By Brian A. Pellerin, Brian A. Bergamaschi, Bryan D. Downing, John Franco Saraceno, Jessica D. Garrett, and Lisa D. Olsen

Abstract

The recent commercial availability of in situ optical sensors, together with new techniques for data collection and analysis, provides the opportunity to monitor a wide range of water-quality constituents on time scales in which environmental conditions actually change. Of particular interest is the application of ultraviolet (UV) photometers for in situ determination of nitrate concentrations in rivers and streams. The variety of UV nitrate sensors currently available differ in several important ways related to instrument design that affect the accuracy of their nitrate concentration measurements in different types of natural waters. This report provides information about selection and use of UV nitrate sensors by the U.S. Geological Survey to facilitate the collection of high-quality data across studies, sites, and instrument types.

For those in need of technical background and information about sensor selection, this report addresses the operating principles, key features and sensor design, sensor characterization techniques and typical interferences, and approaches for sensor deployment. For those needing information about maintaining sensor performance in the field, key sections in this report address maintenance and calibration protocols, quality-assurance techniques, and data formats and reporting. Although the focus of this report is UV nitrate sensors, many of the principles can be applied to other in situ optical sensors for water-quality studies.

Introduction and Background

The recent commercial availability of in situ sensors, together with new techniques for data collection and analysis, provides the opportunity to monitor water quality on the time scales in which changes occur. In particular, optical sensors—those that measure constituents in the environment by their absorbance or fluorescence properties—have had a long history in oceanography for measuring highly resolved concentrations and fluxes of organic matter and nutrients, but are only recently emerging as useful tools for freshwater studies. Optical sensor technology is sufficiently developed to warrant broader application, but collecting data that meet high-quality standards requires investment in and adherence to common methods and protocols for sensor selection, characterization, and operation, as well as for data quality assurance, control, and management (Pellerin and others, 2012).

Of recent interest is the application of ultraviolet (UV) photometers for the in situ determination of nitrate concentrations in surface waters. Nitrate is important because of its roles in regulating plant growth, eutrophication or hypoxia in aquatic systems, and potential human health effects (Vitousek and others, 1997; Townsend and others, 2003; Bryan and Loscalzo, 2011). UV nitrate sensors have been used during the past few decades for wastewater monitoring (Rieger and others, 2008; Drolc and Vrtovšek, 2010) as well as for coastal and oceanographic studies (Johnson and Colletti, 2002; Johnson and others, 2006; Johnson, 2010; Zielinski and others, 2011), but have gained broader use in freshwater systems only in the last few years, from which we have gained an improved understanding of the magnitude and drivers of nitrate variability (Sanford and others, 2007; Pellerin and others, 2009; Heffernan and Cohen, 2010; Pellerin and others, 2011; Cohen and others, 2013).

All the UV nitrate sensors presently available operate on the same basic principle—the absorbance of light by nitrate at a specific wavelength is measured by a photometer and converted to a nitrate concentration. The UV approach offers several advantages compared to wet chemical nitrate sensors and ion-selective electrodes (table 1). However, UV nitrate sensors from individual manufacturers differ in several important ways that affect their ability to accurately measure in situ nitrate concentrations in different systems. This report provides information about instrument differences to facilitate appropriate applications and usage guidelines that help ensure the collection of high-quality data.

Table 1. Advantages and disadvantages of technologies for in situ measurement of nitrate in freshwater systems.

Advantages	Disadvantages
Ultraviolet nitrate sensors; spectral absorption by a spectrophotometer	
· High resolution, accuracy, and precision · Large nitrate range · Chemical-free · Fast response time · Additional optical information in spectra	· Expensive · High power requirement · High maintenance costs · Subject to a range of optical inteferences
Wet-chemical nitrate sensors; wet chemical colorimetric reaction with nitrate, detection by photometry	
· High resolution, accuracy, and precision · Potential for in situ calibrations · Relatively fast response time	· Expensive · High power requirement · High potential for fouling · High maintenance costs · Requires reagents (generates waste)
Ion-selective electrodes for nitrate; direct potentiometry between a sensing electrode and a reference electrode	
· Inexpensive · Easy to use · Fast response time · Large nitrate range · Not influenced by color or turbidity	· Low resolution, accuracy, and precision · Subject to ionic interferences · High instrument drift

Purpose and Scope

The purpose of this report is to provide information on the selection and use of UV nitrate sensors by the U.S. Geological Survey (USGS) for measuring nitrate concentrations in situ in environmental waters. The goal of this report is to help USGS personnel to collect reproducible nitrate-concentration data by using UV sensors in ways that are comparable across studies, sites, and instruments. For those in need of technical background and information about sensor selection, the following topics will be of greatest interest: (1) operating principles, key features, and sensor design; (2) sensor characterization techniques and typical interferences; and (3) approaches for sensor deployment. However, for those in need of information related to maintaining sensor performance and data quality, the following topics will be most relevant: (1) sensor maintenance and calibration protocols, (2) quality-assurance techniques, and (3) appropriate data formats and reporting. This report is not intended to replace the manufacturer's user manuals for individual sensors, nor does it present a comprehensive comparison of all commercially available UV nitrate sensors.

Related Information

This report is designed to supplement other documents describing collection and reporting of water-quality data, including the deployment and use of continuous water-quality monitors, which are presented in order of general to more specific guidelines:

- **The USGS Fundamental Science Practices** (Survey Manual *502.1*, *502.2*, *502.3*, and *502.4*): The Fundamental Science Practices (FSPs) are a collection of policies for ensuring the quality and integrity of USGS science that include procedures for planning and conducting data collection and research (*502.2*), peer review (*502.3*), and review, approval, and release of information products (*502.4*). Core values that pertain to the use of UV nitrate sensors include documenting the "methods or techniques used to collect, process, or analyze data," which comprise the accuracy and precision, standards for metadata, and methods of quality assurance. Plans for data collection are documented in proposals or work plans that are approved at a level higher than the project, and data collection is carried out in a consistent, objective, and replicable manner that has been vetted through a vigorous and open process of peer review. Techniques used by USGS scientists conform to, or reference, national and international standards and protocols, if they exist (such as this report), and are reviewed by a minimum of two qualified peer reviewers before publication in information products. This report facilitates adherence to the FSPs by providing a citable reference for protocols for use and guidelines for documenting the metadata needed to describe the results collected by UV nitrate sensors.
- **The USGS National Field Manual for the Collection of Water-Quality Data**: The National Field Manual (*NFM*) provides information and guidelines on preparing for sampling, selecting and cleaning equipment, collecting and processing water samples, and taking field measurements. Much of the information in the NFM is directly applicable to deploying UV nitrate sensors and collecting the metadata needed for interpreting the results. For example, *Chapter A1* (U.S. Geological Survey, variously dated, preparations for water sampling) includes making checklists of needed supplies and equipment and establishing sampling sites in the National Water Information

System (NWIS). *Chapter A2* (U.S. Geological Survey, variously dated, selection of equipment) includes information on selecting tubing, gloves, blank water, and other supplies in consideration of chemical compatibility with the constituents being measured; for UV nitrate sensors it is important to select supplies that do not leach substances that interfere with UV absorbance in the 200–250 nanometer (nm) range (for example, color, turbidity, or organic matter). *Chapter A3* (U.S. Geological Survey, variously dated, cleaning of equipment) includes supplies and procedures for cleaning inorganic constituent sampling equipment. *Chapter A4* (U.S. Geological Survey, variously dated, collection of water samples) includes methods for collecting water-quality samples, which ensure that the UV nitrate sensor results are representative of the field conditions by comparison to laboratory results, and collecting quality-control samples, which ensure that the UV nitrate sensor results are not substantially affected by measurement bias or variability. *Chapter A6* (U.S. Geological Survey, variously dated) (field measurements) includes methods for making ancillary field measurements that are useful for understanding the UV nitrate sensor results, such as temperature, dissolved oxygen, specific electrical conductance, pH, and turbidity. Personnel who use UV nitrate sensors in USGS studies can benefit from a solid understanding of the guidelines provided in the NFM.

- **Techniques and Methods 1-D3** (*Guidelines and Standard Procedures for Continuous Water-Quality Monitors: Station Operation, Record Computation, and Data Reporting*) by Wagner and others (2006): *TM 1-D3* provides basic guidelines and procedures for use by USGS personnel for site and water-quality monitor selection, field procedures, calibration of continuous water-quality monitors, record computation and review, and data reporting. Although the use of UV nitrate sensors requires specific guidelines beyond those provided in TM 1-D3, the general work flow for the use of UV nitrate sensors is similar to the work flow described for the basic sensors in TM 1-D3. The terminology and processes described in TM 1-D3 these form the general background for these guidelines for the use of UV nitrate sensors.

- **WRD Policy Memorandum 2010.02** on continuous records processing of water time-series data: This document specifies a schedule for the review and approval of time-series data collected in support of USGS activities. Results from most UV nitrate sensors generally are expected to meet the criteria for "Category 2," which covers sites that require seasonal data (rather than shorter periods) for record computation, and includes sites with continuous water-quality analyzers that depend on laboratory results for verification. Category 2 time-series records are finalized within 240 days of collection. If site-specific issues prevent the timely review and approval of the data, then the site could be designated "Category 3" (where continuous record processing does not apply); however, these types of sites are to be rare. A Category 3 designation can be appropriate for sites that require complex data modeling to "fit" the raw UV nitrate sensor output to a series of laboratory or field measurements collected over a wide range of hydrologic and environmental conditions; in these cases, the ancillary data and models used to obtain the final nitrate results can be documented in an interpretive publication that undergoes USGS technical review.

Evaluating the Need for Continuous Data

Given the current costs to purchase a UV nitrate sensor ($15,000–$25,000 per unit) and the ongoing expenses related to instrument service and maintenance, potential users could want to carefully consider whether "continuous" nitrate data (for example, multiple samples per day) are really needed. Although explicit guidelines are not available, basic time-series analysis requires that the rate of sampling be greater than the rate of change to observe the true time-dependence. Sampling bias or aliasing can occur when constituent concentrations change significantly between samples, which, in turn, can lead to over-estimates or underestimates of watershed loads, inaccurate pollution assessments, and potentially obscured seasonal or long-term trends. Traditional discrete sampling approaches that result in 12–18 samples per year can be particularly susceptible to aliasing problems in dynamic freshwater and coastal systems.

There are many examples of locations or studies where frequent data are critical for understanding drivers of water quality and resultant effects on human health, ecosystem function, or water management. For example, continuous measurements, in some cases, can improve the calculation of nitrate trends and loads, where discrete sampling cannot fully represent the concentration-discharge relationship or where the concentration-discharge relationship is poor. Similarly, continuous measurements are often critical for developing process-level understanding of sources and alteration of nutrients. Nevertheless, not all systems are subject to rapid changes in nitrate concentration, and some questions are addressed sufficiently with less frequent, discrete data collection. Potential users can evaluate existing discrete data and continuous sensor data for other parameters (such as specific conductance and dissolved oxygen) to determine if a site would benefit from continuous in situ measurements. If sufficient information is not available at a given site, temporary sensor deployments could provide short-term, but useful, data on the degree of variability before investing in a permanent continuous measurement effort.

Principles of UV Absorbance Measurements for Nitrate

Absorbance is a dimensionless measurement of the ability of a medium (for example, a water sample) to absorb light (Hu and others, 2002). Absorbance (A) occurs when a photon emitted from a light source excites an electron from a ground state to a higher energy orbital, and it is represented by the following equation:

$$A_\lambda = -\log\left(\frac{I_\lambda}{I_{\lambda,o}}\right) \quad (1)$$

where

A_λ is the absorbance at a specific wavelength (λ),
I_λ is the intensity of the light at wavelength λ passing through a sample, and
$I_{\lambda,o}$ is the intensity of incident light at wavelength λ before it enters the sample.

The concentration of an absorbing substance in solution can be calculated from absorbance measurements according to Beer's Law, if the amount of light absorbed per molecule of the substance of interest and the amount of sample through which the light travels are known (as in fig. 1) on the basis of the following equation:

$$c = \frac{A_\lambda}{\varepsilon_\lambda * L} \quad (2)$$

where

c is the concentration of the absorbing substance,
A_λ is the absorbance at a specific wavelength (λ),
ε_λ is the molar absorptivity of the absorbing substance at wavelength λ (a constant), and
L is the path length.

Although absorbance is the property used to calculate concentration, a photometer does not technically measure absorbance; it measures the amount of incident light at a given wavelength ($I_{\lambda,o}$) that is transmitted through the solution to the detector (that is, transmittance). Absorbance and transmittance are related logarithmically (fig. 2), as described by the following equation:

$$A = 2 - \log_{10} \%T \quad (3)$$

where

A is absorbance, and
$\%T$ is transmittance as a percentage (that is, 100 x I/I_o).

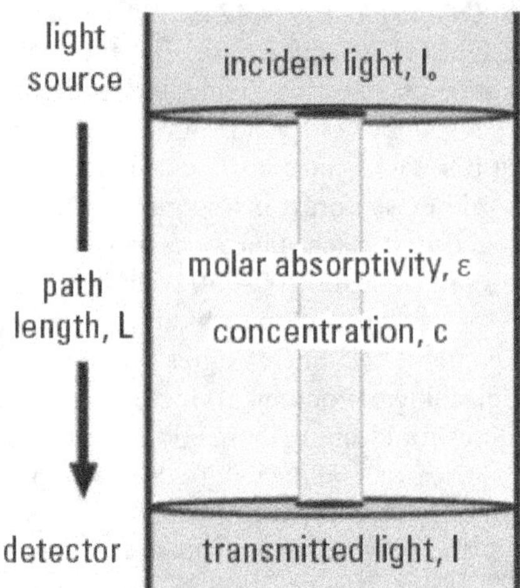

Figure 1. Conceptual diagram demonstrating the attenuation of a beam of radiation by an absorbing solution in a typical ultraviolet (UV) nitrate-sensor flow path.

A solution with no absorbing substances will transmit all incident light to the detector, thereby yielding a transmittance of 100 percent and an absorbance value of 0 absorbance units (AU). Samples with a high concentration of absorbing substances can transmit only a small fraction of incident light through a sample, and, for example, could yield a transmittance of 0.01 percent at an absorbance value of about 4 AU (fig. 2). As will be discussed, reduced transmittance of light due to substances other than nitrate is one of the considerable challenges for using UV nitrate sensors in natural waters.

The ability to measure nitrate concentrations on the basis of UV absorption measurements with laboratory photometers was demonstrated more than 50 years ago (Bastin and others, 1957; Armstrong, 1963) and is a standard method for nitrate screening in samples with low organic matter and particle interference (Standard Method 4500-NO_3- B. Ultraviolet Spectrophotometeric Screening Method; American Public Health Association, American Water Works Association, and Water Environment Federation, 1995). In comparison to standard wet chemical methods for nitrate, such as cadmium reduction, the UV absorption method has the advantage of being simple, inexpensive, and chemical-free. The measurement is a direct spectrophotometric determination of nitrate concentration, where the magnitude of the absorbance at a peak wavelength of 220 nanometers (nm) is proportional to the concentration of nitrate ions in solution (fig. 3*A*); however, the measurement is subject to interferences from inorganic and organic substances that reduce the transmission of light (that is, attenuation) at wavelengths similar to nitrate, including nitrite, bromide, chromophoric dissolved organic carbon (DOC), and turbidity (fig. 3*B*). Accurate determination of nitrate concentrations, therefore, requires that light attenuation by interfering substances is negligible or is removed from the absorbance value used to calculate nitrate.

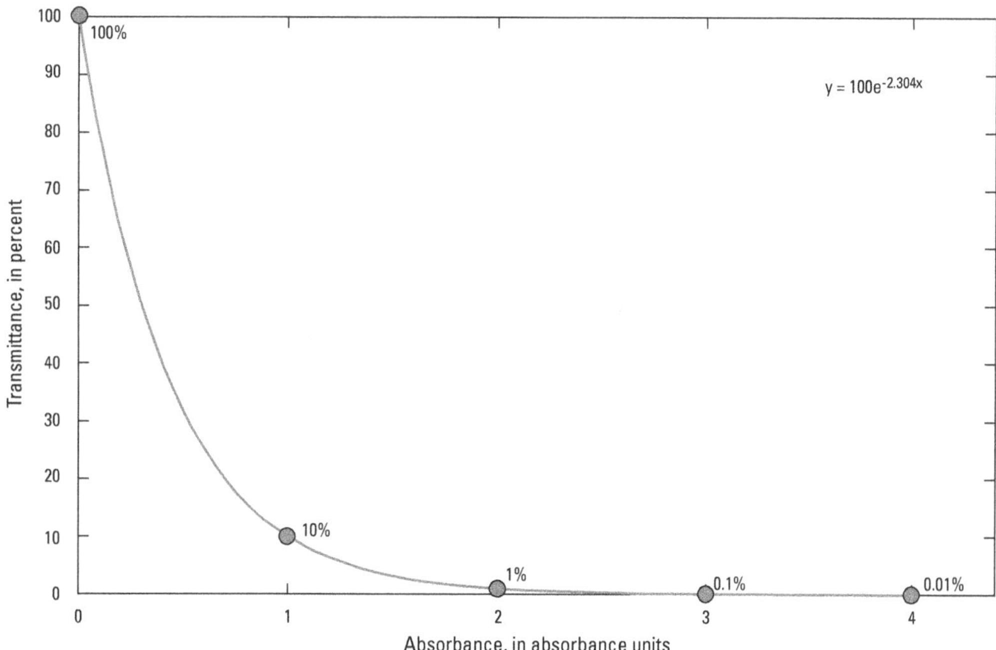

Figure 2. Exponential relationship between absorbance (in absorbance units, AU) and transmittance (in percent, %).

Sensor Design

The primary elements of a field photometer—a regulated light source, a sample path, and a light meter—are fundamentally the same as in a benchtop laboratory photometer. A number of important modifications are needed, however, when a laboratory instrument is miniaturized and adapted for field deployments in harsh and remote environments. These modifications include rugged housings and components, efficient power and heat handling, few moving parts, internal dataloggers and controllers, anti-fouling components, and integrated data processing. These elements not only affect the ability to take accurate nitrate measurements in real-time, but also affect the serviceability, longevity, stability, and cost of field photometers.

In situ UV photometers generally work as follows (fig. 4): a lamp in one end of the housing generates a focused beam of UV light that is directed through a sample and into the detector. The light is reduced (attenuated) as it passes through the sample by constituents in the water sample, and the light reaching the detector (that is, transmitted light) is measured at specific wavelengths in the UV range, which is output as voltage. After additional filtering of the electronic signal and integrated processing to remove interferences, nitrate concentrations and other data are transmitted in analog or digital format (for example, RS-232, SDI-12, or RS-485) to a data-collection platform (DCP) for logging or transmission.

Instrument Components

The overall design of commercial UV nitrate sensors generally is similar among manufacturers and includes the following: (1) physical configuration features, (2) optical features, and (3) data-processing features. However, the engineering design of each type of commercial UV nitrate sensor creates characteristic functional limitations on the types of measurements that can be made and the environments in which they are best deployed. For example, instruments capable of deployment in wastewater treatment facilities are designed for high suspended-particle concentrations (for example, sludge), high fouling rates, and high nitrate concentrations. In contrast, sensors developed for applications in coastal oceans are optimized for clear waters with little biological fouling and low nitrate concentrations. Therefore, choosing the appropriate sensors for different environments requires a clear understanding of the key features and data specifications for the different UV sensors (table 2).

Physical configuration features: **Instrument housings** are made from a variety of materials (for example, acetal, stainless steel, or titanium) that affect their resistance to corrosion and fouling, as well as the pressure and temperature rating. The **path length** is the precise distance of the optical path between the light source and detector windows and varies (less than 1 to greater than 100 millimeters) among manufacturers and models. The path length is a critical feature

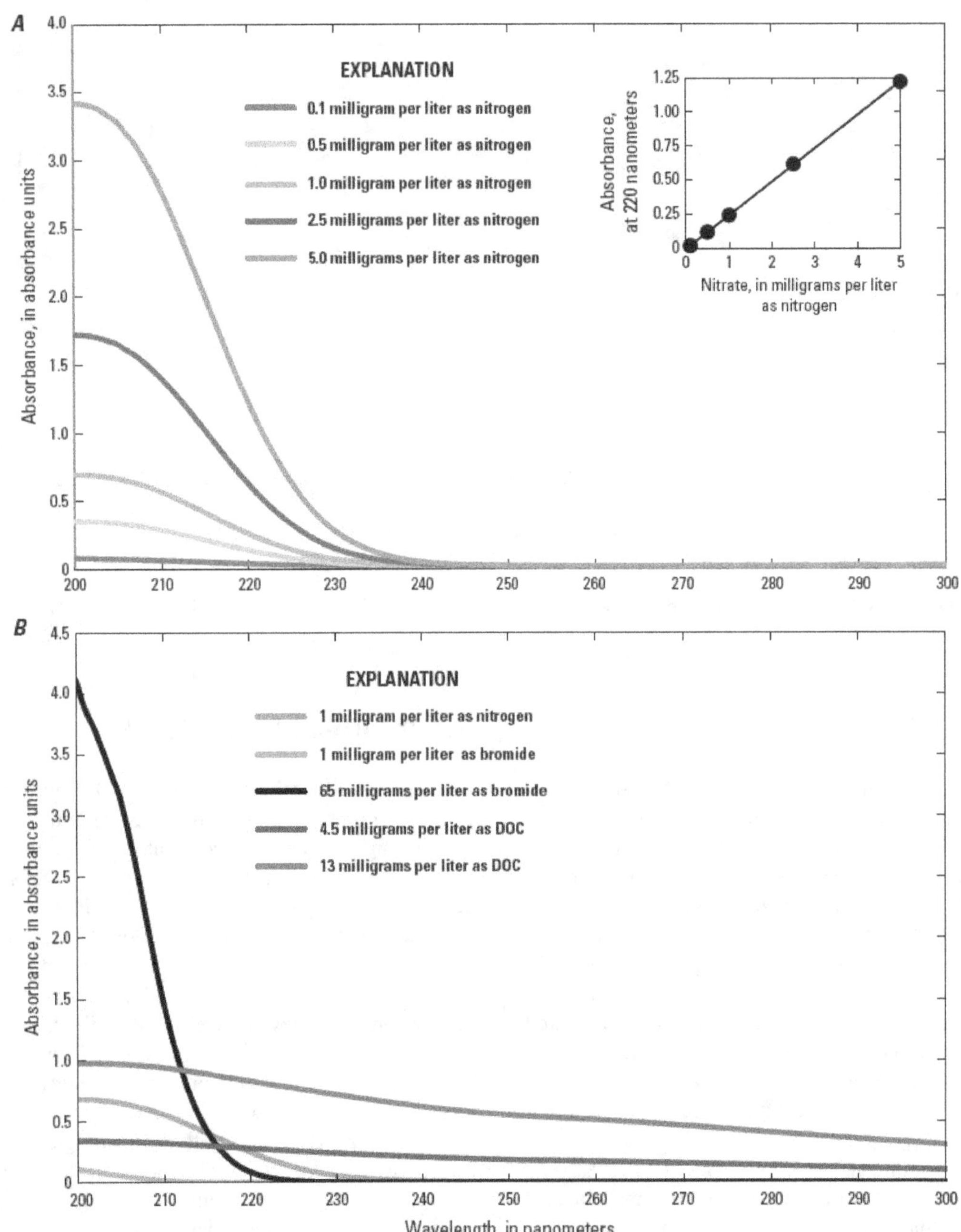

Figure 3. Ultraviolet (UV) absorbance spectra *A*, for nitrate across concentrations ranging from 0.1 to 5 milligrams per liter (mg/L) as nitrogen (N) showing the linearity at 220 nanometers (inset), and *B*, for bromide and dissolved organic carbon (DOC; Suwanee River natural organic matter standard) with 1 mg/L as N nitrate solution.

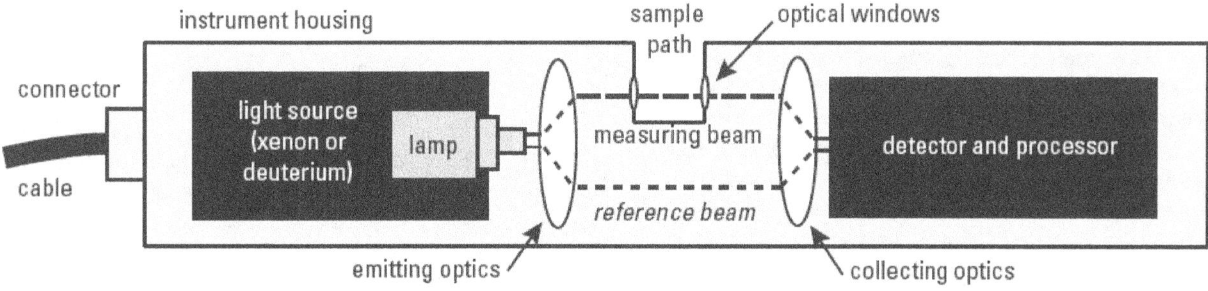

Figure 4. General design and key components of field-deployable ultraviolet (UV) nitrate sensors (modified from Langergraber and others, 2004).

Table 2. Ultraviolet (UV) nitrate sensor design and manufacturer-stated data specifications.

[**Abbreviations**: in, inches; lbs, pounds; m, meters; mg/L as N, milligrams per liter as nitrogen; mm, millimeters; nm, nanometers; sec, seconds; C, degrees Celsius; %, percent]

Parameter	HACH Nitratax	Satlantic SUNA	S::CAN spectrolyzer	TriOS ProPS
Pathlengths available (mm)	1, 2, 5 (fixed)	5, 10 (fixed)	0.5–100 (semi-fixed)	1–60 (semi-fixed)
Wavelengths measured (nm)	220, 350	190–370	200–750	190–360
Approximate dimensions (in)	13.0 x 3.0	21.0 x 2.3	21.5 x 1.7	20.5 x 2.7
Weight in air (lbs)	7.3–7.9	5.4	7.5	11
Housing materials available	stainless steel	acetal, titanium	stainless steel	stainless steel, titanium
Lamp type	xenon	deuterium	xenon	deuterium
Reference beam	yes	no	yes	No
Windows	quartz	quartz	sapphire, fused silica	fused silica + nano coating
Communications	Modbus (RS485, RS232), analog	USB, RS232, SDI-12, analog	Modbus (RS485, RS232), SDI-12, analog	RS232
Power consumption	24 VDC	8…18 VDC	11…15 VDC	9…36 VDC
Connectors	integrated cables	wet pluggable	integrated cables	wet pluggable
Anti-fouling method	wiper (silicone)	wiper (nylon brush)	wiper or compressed air	compressed air + nano coating
Operating Temperature (°C)	2–40	0–40	0–45	0–40
Maximum operating depth (m)	5	100[b]	100	500[b]
Lower detection limit (mg/L as N)	0.1–1.0[a]	0.007	0.03	0.005–0.3[a]
Upper detection limit (mg/L as N)	20–100[a]	28-56[a]	10–70[a]	8.3–500[a]
Accuracy	±3–5% of reading or ±0.5–1.0 mg/L, whichever is greater[a]	±10% of reading or ±0.03-0.06 mg/L, whichever is greater[a]	±2% of reading plus 1/optical path length (in mm; mg/L)	±2% of reading or ±0.155 mg/L, whichever is greater
Precision (mg/L as N)	0.1–0.5[a]	0.028	0.02–0.1[a]	0.03
Maximum sampling interval (sec)	60	1	60	120

[a]Actual specifications dependent on the model used, pathlength, or both.

[b]Options available for deep sea deployments (500 and 2,000 m for SUNA, 6,000 m for TriOS).

of the instruments because absorption by nitrate (as well as effects of interferences) increases linearly with path length. **Power consumption** of these instruments ranges from 2 to 7.5 watts (W) (nominal at 12 volts direct current, VDC), but the total power consumption of the overall system increases with the need for ancillary components, such as pumps (for flow through designs) and anti-fouling measures. **Power cables and connectors** are integrated or replaceable, wet-pluggable components that come in a variety of materials and configurations that affect their strength, durability, and corrosion-resistance. **Dataloggers and controllers** can be integrated or external to the sensor housing and use proprietary or generic data transfer protocols, on the basis of manufacturer, to control the instruments and collect and transmit data. **Anti-fouling measures** include removable or fixed copper components, such as biofouling guards, wipers with nylon bristles or silicon blades (integrated or externally mounted on the sensor), automatic air cleaning systems, or flow cells for integrating pumps and filters.

Optical features: The **light source** in the current generation of optical nitrate sensors is either a xenon flash lamp or a continuous deuterium lamp, which differ in terms of thermal stability, spectral stability, spectral output, brightness, lamp lifetime, failure mode, and power requirements (Finch and others, 1998). The full lifetime of a deuterium lamp is about 1,000 hours, whereas the xenon flash lamps are rated for 2,000-3,000 hours. However, lamp degradation can become evident at half of the lamp lifetime. **Detectors** are the true "sensing" elements of the instrument and, typically, are photodiodes or diode arrays that are more sensitive and stable in the UV region than are other solid state detectors (Johnson and Coletti, 2002). The current generation of sensors differs in terms of the range and number of wavelengths measured by the detector, with some instruments making more than 200 measurements across the full spectral range (about 200–750 nm) compared to one that makes measurements at only two wavelengths in the UV region to account for nitrate (about 220 nm) and interferences from dissolved constituents and particles (about 350 nm). **A reference detector** also can be incorporated into some instruments (sometimes referred to as a "dual beam") to correct for variation in lamp output.

Data-processing features: An **onboard, or ancillary, processor** converts the raw signal to a calculated nitrate concentration on the basis of an instrument-specific algorithm. The purpose of the **algorithm** is to differentiate the nitrate signal from the interfering species, to account for signal loss due to the presence of sediment in the light path, and to remove noise from the measurement. The processor interacts with the data collection platform (DCP) to output the data according to a common transfer protocol (for example, RS-232, SDI-12, or RS-485) to a data logger.

Data Specifications

The current generation of UV nitrate sensors is diverse in terms of data specifications, such as detection limit, measurement range, resolution, accuracy, and precision (table 2). These differences are largely related to the intended original application for the instrument, which can require design tradeoffs to work in a specific type of environment. For example, sensors developed for wastewater often have lower precision and accuracy than those developed for coastal applications, but they can make measurements of nitrate over a much wider concentration range and have a greater tolerance for interferences due to shorter path lengths. Clearly, establishing the data-quality needs for a given study is critical to choosing the most appropriate sensor.

In theory, the sensitivity, accuracy, and concentration range of a field photometer are a trade-off, dependent largely on the path length of the instrument. As described by Beer's Law (eqn. 2) and illustrated in figure 5, increasing the path length increases the number of absorbing molecules in the light path, and results in an exponential decrease in the light transmitted to a detector. If the path length is so long that it permits sediment or molecules in solution to absorb all the light, no measurement can be made because no light reaches the detector. In practice, the useful range of the instrument is typically determined by the path length, whereas the accuracy is determined by the path length, the instrument noise (derived from fluctuations in the lamp, detector, and associated circuitry), and the matrix effects.

Although manufacturers provide data-quality specifications, the true specifications of the instrument can be higher or lower than stated (fig. 6). In particular, manufacturer specifications are often determined by using nitrate standards in ultra-pure water, which would typically result in greater

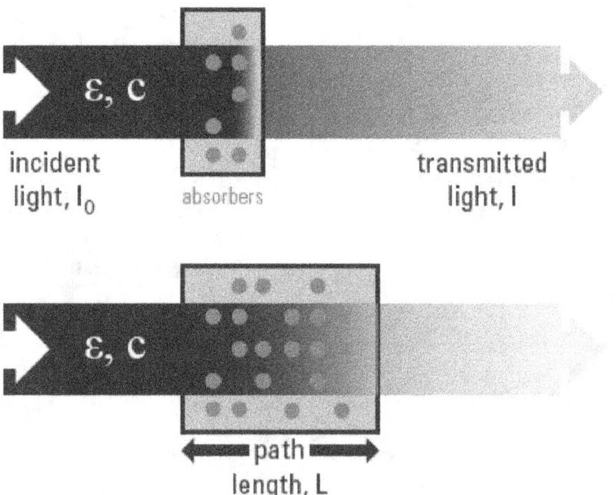

Figure 5. Conceptual diagram showing the effect of a shorter (top) and longer (bottom) optical path length (L) on the amount of light transmitted (I), when the molar absorptivity (ε) and concentration (c) of absorbers are constant.

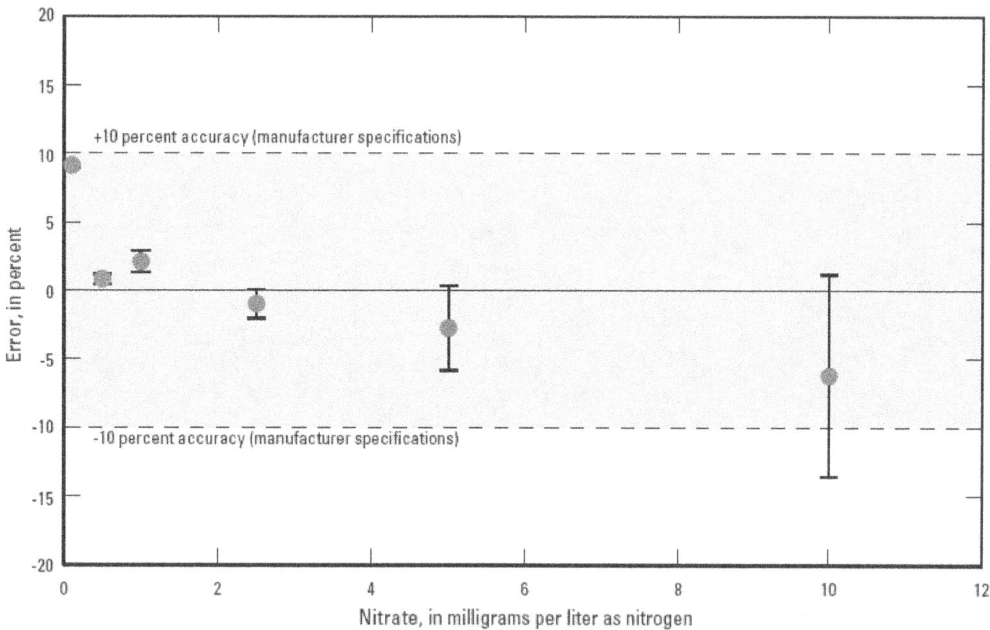

Figure 6. Measured mean error and standard deviation in nitrate concentrations under laboratory conditions with nitrate standards in ultra-pure water in comparison to manufacturer-stated error specifications.

accuracy, greater precision, and lower detection limits than if measured in natural waters where matrix interferences are a factor. Therefore, it is important that the true performance of an instrument is verified in the laboratory with nitrate standards and matrix spikes prior to initial deployment and after instrument servicing. Guidelines for evaluating the instrument accuracy, precision, detection limit and linearity are included in appendix 1. The following, however, are typical instrument specifications provided by the manufacturer.

The measurement range is defined as the difference between the greatest and least measurable values. Together, the manufacturer-reported ranges for the current generation of UV nitrate sensors span the range of concentrations reported in virtually all surface and ground waters (0–100 mg/L as nitrogen, or N), but no individual instrument spans the full range. The primary factor determining the measurable range of nitrate concentrations is the optical path length—the path length must be short enough for adequate light to reach the detector, but long enough for a measurable difference between the incident and transmitted light. Sensors with longer path lengths typically have a lower measurable range of nitrate concentration, whereas sensors with shorter path length allow for measurements over a greater range of concentrations (fig. 7). Currently, instruments are available with path lengths of less than 1 mm to more than 100 mm, with 2–10 mm being most common path length in instruments used for freshwater and coastal deployments. Longer path lengths can be used for clear-water applications, such as drinking water, whereas shorter path lengths (less than 2 mm) are more commonly used for settings with high nitrate concentrations (for example, greater than 20 mg/L as N), high dissolved organic carbon (DOC) concentrations, or high suspended particle concentrations, such as wastewater-treatment plants.

The detection limit is defined as the lowest value measurable by a given sensor at the 99-percent confidence level. Detection limit, like range, is mostly affected by the path length, but is also a function of the detector sensitivity and the instrument electronic noise. The detection limit of the current generation of UV nitrate sensors ranges from less than 0.01 to 1.0 mg/L as N (table 2) and can be an important consideration when selecting a sensor for a particular study. Repeated measurements or instrument-specific calibrations, in some cases, can be used to determine a true detection threshold above or below that reported by the manufacturer (see appendix 1).

Resolution is defined as the minimum difference between measured values reliably detected by the sensor. Resolution is a critical, but often overlooked, characteristic of instruments because it defines the minimum reportable variation in nitrate concentrations. The accuracy of an instrument can be no greater than the resolution. Figure 8 illustrates the effect of instrument resolution on time-series measurements with the same data shown at one and two significant figures. Note that the lower resolution data indicates periods of apparent noise when concentrations are fluctuating rapidly.

Accuracy is the degree of agreement between the measured nitrate value and its true quantity. Accuracy is usually calculated by comparison to known concentrations in standard solutions in ultra-pure or matrix water. Manufacturer stated accuracies are typically reported as ±3–10 percent of the measured nitrate concentration or as a fixed nitrate concentration

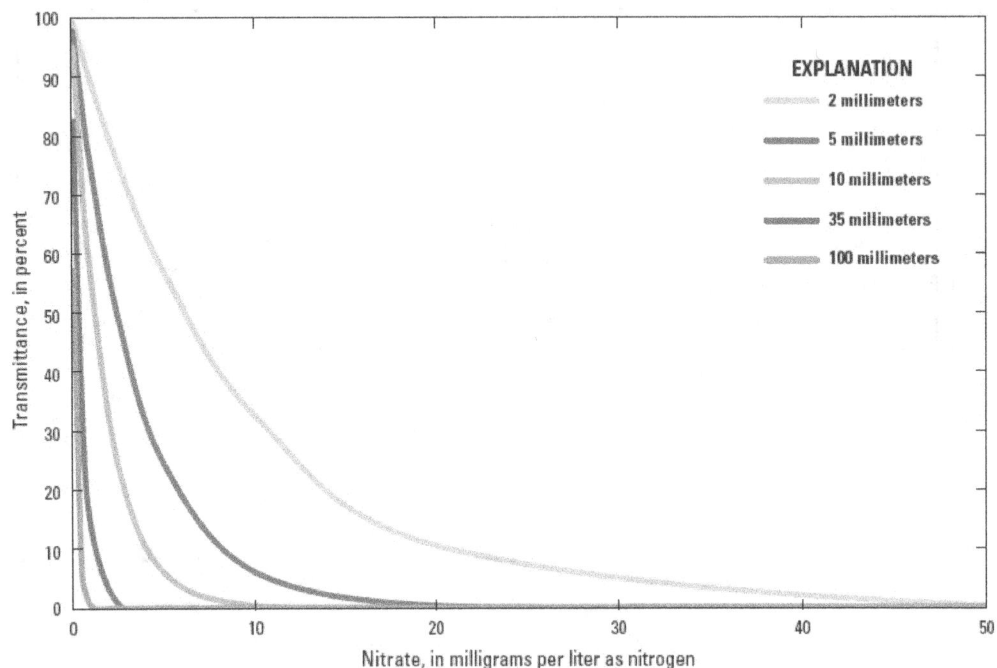

Figure 7. Nitrate concentration compared to transmittance for sensors with different theoretical path lengths (2–100 millimeters, mm). Calculations are based on Beer's Law and use a molar (M) extinction coefficient (ε) for nitrate at 220 nanometers of 3,449 M per centimeter.

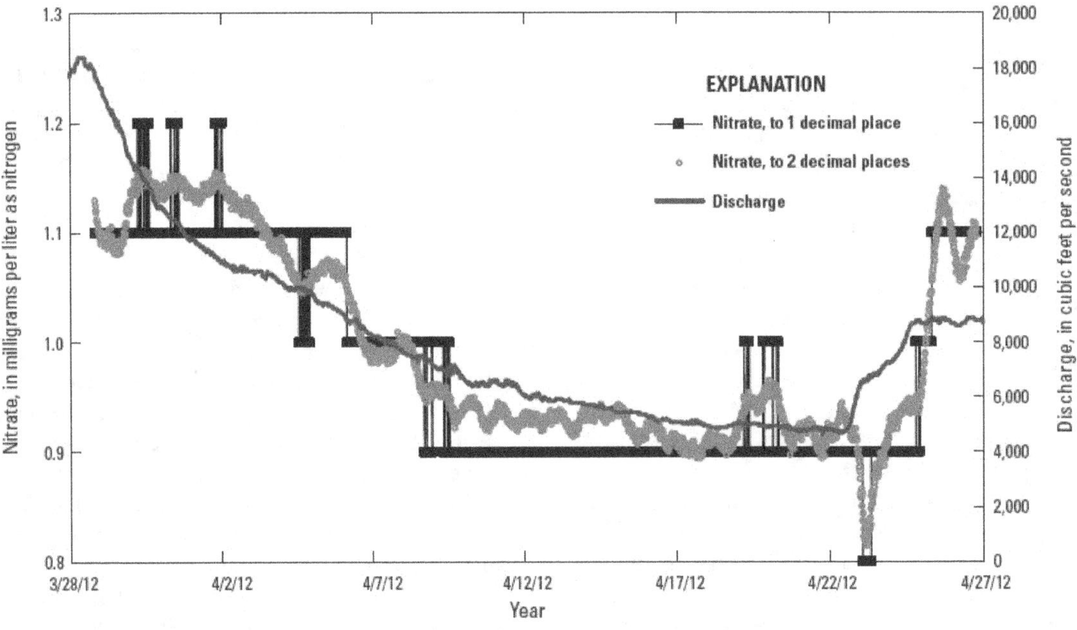

Figure 8. Nitrate concentration relative to discharge from the Potomac River at Little Falls, Maryland (U.S. Geological Survey gage #01646500), collected by using a Satlantic SUNA sensor that has a resolution of two significant figures (red circles). The black squares and line are the same data rounded to one significant figure.

(whichever is greater). As such, the relative accuracy of UV nitrate sensors inherently decreases at low nitrate concentrations when accuracy is dictated by the fixed concentrations (which typically range from 0.03 to 1.0 mg/L as N). As noted previously, however, manufacturer specifications are often determined by using nitrate standards in ultra-pure water, which typically results in greater accuracy, greater precision, and lower detection limits than if measured in natural waters where matrix interferences are a factor. Therefore, it is important that the true performance of an instrument is verified in the laboratory with nitrate standards and matrix spikes prior to initial deployment and after instrument servicing.

The accuracy of UV nitrate sensors is primarily affected by two important sources of uncertainty: (1) instrument noise and (2) matrix effects. Although instrument noise, due to the electronic components, the lamp, and the photometer resolution, is often random (Skoog, 1985), a systematic error can result in good precision but poor accuracy (that is, bias). Instruments with a reference channel that can account for some changes or noise in lamp output, or those that integrate multiple scans in a single measurement, typically have greater accuracy. In addition, matrix effects, such as high concentrations of suspended particles or dissolved organic matter, can also decrease accuracy through decreasing transmittance or increasing error at wavelengths used to calculate nitrate concentrations. Instruments that use the full spectrum or multiple wavelengths within the spectrum tend to be more accurate in the presence of interfering substances because the algorithms are typically better able to account for the interferences.

Precision is the range of values reported by the sensor when making repeated measurements of the same sample under the same conditions. As with other data specifications, manufacturer-stated instrument precision is typically determined for measurements of relatively high nitrate concentrations or in the absence of matrix interferences. In practice, instrument precision can degrade dramatically at low nitrate concentrations, in the presence of optical interferences (see "Matrix Effects" section), and under some environmental conditions. Adequate performance of the instrument under field conditions at the lowest anticipated nitrate concentration is necessary to ensure the data meet quality-assurance criteria of the application.

Matrix Effects

To accurately measure nitrate optically in natural waters, it is critical to account for light-absorbing or light-scattering materials present in the sample that interfere with light transmission to a detector. Collectively, these are known as "matrix effects" because they result from properties of the matrix in which the measurement of nitrate is being made. Differences in instrument design (path length, lamp output, and detector wavelengths) and in spectral processing algorithms are key to correcting for particular interferences. Therefore, evaluating the type, magnitude, and presence of interfering substances—as well as the ability to correct for systematic error associated with matrix effects—is critical for good sensor selection and for assuring data quality.

The two principal matrix effects—those from dissolved substances and suspended particles—are discussed in detail in the following sections. Other potential interferences, such as microalgae, air bubbles, and direct sunlight, are typically addressed with mechanical solutions, such as wipers and shade caps, and are not addressed in detail in this report. In addition, alternative strategies for reducing matrix effects could be necessary in especially challenging environments. For example, it is possible to filter out particles prior to the nitrate measurement with an in situ apparatus, but complexity, filter costs, and frequent site visits make this approach impractical in many settings. Specific remedies, such as this, however, are not discussed in this report because they are relatively uncommon and are often application-specific.

Absorbance by Dissolved Constituents

A number of dissolved constituents absorb light in the UV wavelength range used to calculate nitrate concentrations. These include inorganic constituents, such as bromide, hydrogen sulfide, and nitrite, as well as colored dissolved organic matter, such as humic and fulvic acids. The presence of these constituents reduces the transmittance of light through a sample and can result in an overestimate of nitrate if not accounted for. Because the shape of an absorbance spectrum varies by constituent, several UV nitrate sensor manufacturers use an absorption curve-fitting technique and laboratory-calibrated extinction coefficients to account for overlapping absorbance from interfering species. Most instruments use multiple wavelengths to distinguish the absorption due to nitrate from that due to other substances in the matrix, but one sensor design uses a single wavelength (350 nm) to simultaneously account for all non-nitrate interferences (for example, particles and organic matter). The effect of interfering substances on the final reported nitrate value can be significant, even when using instruments that have integrated compensation techniques. For example, a positive bias was observed in the reported nitrate value with increasing dissolved organic carbon (DOC) concentrations during a comparison of two types of UV nitrate sensors in a laboratory setting (fig. 9). The increase was particularly noticeable in the data from the dual wavelength, 2-mm path-length instrument for waters high in organic matter (Drolc and Vrtovšek, 2010).

Note that the algorithms used to calculate nitrate concentrations not only differ from manufacturer to manufacturer, but also by intended application for a specific instrument. This is a particularly important consideration for manufacturers whose instruments are used in a variety of complex matrix types ranging from drinking water to wastewater. For example, data collected by using the full spectrum, 5-mm path length sensor during the laboratory DOC additions shown in figure 9

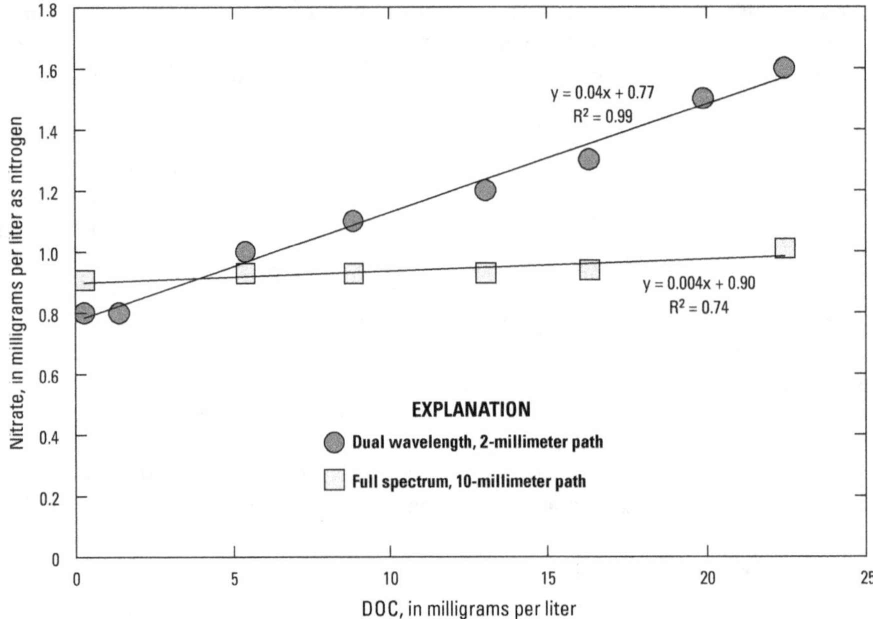

Figure 9. Reported nitrate concentrations at a range of dissolved organic carbon (DOC) concentrations in a laboratory study comparing a two-wavelength sensor (HACH Nitratax, 2-millimeter path length) and a full ultraviolet (UV) spectrum sensor (Satlantic SUNA, 10-millimeter path length). Suwanee River natural organic matter (1R101N; International Humic Substances Society) was added as the DOC source to a 1 milligram per liter as nitrogen nitrate standard solution.

and processed by using three different manufacturer "calibrations" showed overestimates of nitrate concentrations when interferences were not adequately accounted for (fig. 10). This illustrates the importance of using the appropriate matrix compensation algorithms when calculating nitrate concentrations in natural waters.

When deploying instruments in estuarine and coastal settings, interferences from bromide also can be significant, if not accounted for by the instrument. At typical sea water concentrations (about 65 mg/L bromide at a salinity of 35 practical salinity units, or PSU), absorbance by bromide at wavelengths around 190–220 nm could have a strong effect on calculated nitrate concentrations. This is particularly important because absorbance by bromide is temperature dependent (Sakamoto and others, 2009; see more on temperature effects later in this report).

Results from a laboratory bromide addition showed the effect of high bromide concentrations on nitrate measurements with a two-wavelength sensor that does not compensate for interferences by bromide (fig. 11). In contrast, full UV-spectrum sensors typically measure the wavelengths necessary to make a correction, but can require the use of a separate seawater compensation algorithm (fig. 11). However, the bromide concentration in freshwater environments typically is less than 1 mg/L, indicating that the interference would be negligible in most rivers, streams, and groundwater systems.

In limited circumstances, other dissolved constituents, including nitrite, sulfide, iron, and iodine, can be important interferences on UV nitrate sensor measurements. However, the concentration in freshwaters typically is low enough to minimize any interference on nitrate measurements. If dissolved constituents are present at high concentrations at the intended study site, instrument performance in the presence of these interfering species can be assessed prior to deployment.

Scattering by Particles

Scattering of light by suspended material in the optical path reduces the light reaching the detector and, therefore, can result in an overestimate of sample absorbance (Roesler, 1998). Scattering by inorganic particles is generally assumed to be uniform across the UV and visible range and, therefore, is unlikely to affect nitrate concentrations calculated from the shape of the absorption curve rather than the absolute magnitude. At high suspended-particle concentrations, however, the signal-to-noise ratio decreases, and transmittance ultimately approaches zero. In addition, organic particles have a spectrally varying signal (Sathyendranath and others, 1987) that potentially could affect both the magnitude and shape of the spectra and, in turn, the calculated nitrate concentrations.

The effects of particles on nitrate calculations can be significant and varies between instrument types. A lab comparison of UV nitrate sensors in a solution of varying inorganic sediment concentrations showed the tendency to overestimate nitrate concentrations at high turbidity (fig. 12). However, the instrument with the longer path length (10 mm) did not report a value at turbidities above about 500 nephelometric turbidity

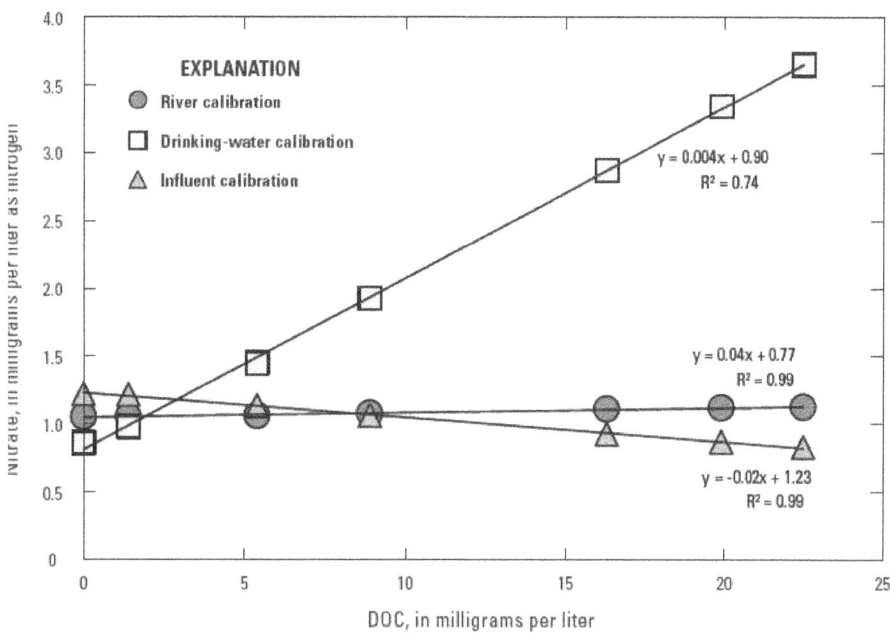

Figure 10. Reported nitrate concentrations at a range of dissolved organic carbon (DOC) concentrations in a laboratory study using a full ultraviolet/visible light spectrum sensor (s::can spectrolyzer, 5-millimeter path length) and processed by using three different manufacturer-defined calibration algorithms. Suwanee River natural organic matter (1R101N; International Humic Substances Society) was added as the DOC source to a 1 milligram per liter as nitrogen nitrate standard solution.

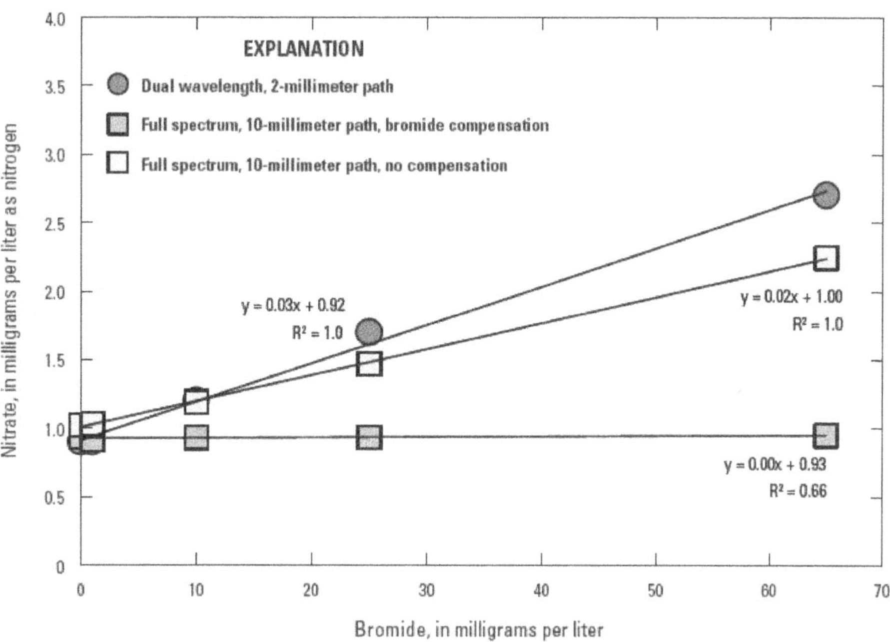

Figure 11. Reported nitrate concentrations in standards with a range of bromide concentrations in a laboratory study using a two wavelength sensor (HACH Nitratax, 2-millimeter path length) and a full ultraviolet (UV) spectrum sensor (Satlantic SUNA, 10-millimeter path length). Bromide was added to a 1.0 milligram per liter as nitrogen nitrate standard solution at constant temperature. The SUNA nitrate concentration was calculated with and without bromide compensation.

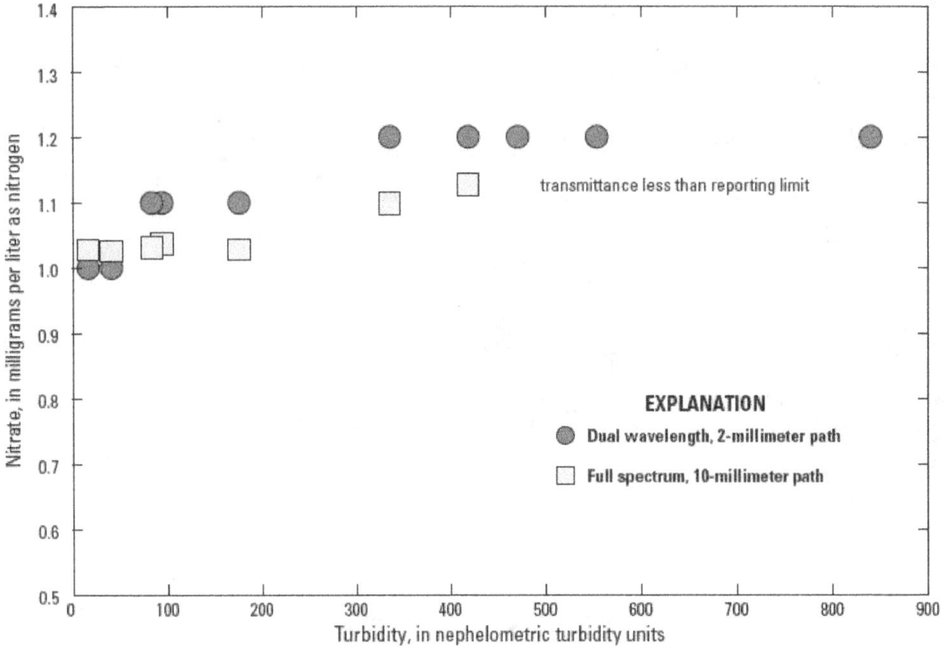

Figure 12. Example of the effect of turbidity (suspended Elliot silt loam soil from IHSS) on nitrate concentrations from a two wavelength (Hach Nitratax, 2-millimeter path length) and full ultraviolet (UV) spectrum (Satlantic SUNA, 10-millimeter path length) sensor. Actual nitrate concentration was 1 milligram per liter (mg/L) as nitrogen (N). For reference, the addition of 4 grams per liter of Elliot silt loam soil resulted in a turbidity of about 1,000 nephelometric turbidity units (NTU). Transmittance was less than the reporting limit at approximately 450 NTU for the 10-millimeter full spectrum sensor, and data are therefore not shown.

units (NTU), which presumably is the point at which insufficient light was reaching the detector. The results shown in figure 12 are for illustrative purposes only, however, because the absolute turbidities at which a UV nitrate sensor can operate depend on the turbidity sensor being used and the type of interfering particles (that is, inorganic versus organic).

Temperature Effects

Previous studies have shown that UV absorbance by nitrate is not temperature dependent, and temperature generally has little influence on the shape of the absorption spectra (Sakamoto and others, 2009). However, temperature does have an effect on bromide absorbance coefficients, and it is compensated for in some instruments that make full UV spectrum measurements (Sakamoto and others, 2009). Although nitrate sensors are particularly susceptible to internal heating, given the use of high-energy lamps, the effect of internal temperature on the measured nitrate concentrations in standard solutions across a range of about 0 to 30 degrees Celsius (°C) is typically very small (that is, less than 0.003 mg/L as N for each degree C; Pellerin and others, 2011). Therefore, the overall effect of temperature on nitrate sensor measurements is likely to be negligible in most freshwater systems when instruments are run for seconds to minutes per interval.

Sensor Selection

A number of factors come into play when selecting the appropriate UV nitrate sensor for use in field settings. Individual sensor selection can be determined by the expected range of environmental conditions, data-quality specifications, and logistical constraints. Differences in sensor design, such as path lengths and the wavelengths measured, are critical features that affect data quality and, consequently, are important to be considered along with the depth rating, temperature rating, and maintenance requirements. Key questions to consider when selecting a UV nitrate sensor for field deployment include the following:

1. **What is the expected range of environmental conditions at the site?**
    - What are the expected ranges in concentration of interfering constituents, including suspended sediment, dissolved organic matter, and bromide?
    - What are the expected ranges of temperature and maximum depth where the instrument will be deployed?
    - What is the expected level of biological or mineral fouling?

2. **What are the data specifications for nitrate concentrations at the site?**
    - What is the expected range in nitrate concentrations?
    - What accuracy, precision, and detection limit are needed, given the study goals?
    - What is the optimal sampling interval, and how many measurements are needed, per interval, to achieve study goals?
3. **What are the site requirements and logistical considerations?**
    - Will instruments be accessible by land or by boat?
    - Will the instruments be accessible across the range of hydrologic and weather conditions expected for the site (including ice cover)?
    - Does the site have existing infrastructure, power, and communication systems?
    - What is the anticipated frequency of site visits?
    - What level of technical expertise is available to manage the instruments?

Many of these questions can be answered with historical water-quality data, site documentation, and data-quality objectives determined for the study. The user can then evaluate instrument needs, relative to manufacturer stated specifications (table 2), and relevant publications. To address matrix effects and other challenges for which data are not readily available, example procedures are included in table 3.

Tradeoffs often are a consideration when selecting a UV nitrate sensor for a specific study. For example, shorter path-length sensors would tend to work best at higher turbidities, but also result in lower accuracy and detection limits for nitrate concentrations. Similarly, sensors with proprietary controllers are easier to use but, ultimately, can limit user flexibility in the type and frequency of data collection. In addition, at some locations, continuous UV nitrate measurements would not be possible without modification to the procedures and protocols described in this report or to the instruments themselves. In the case that modifications are necessary, careful documentation of the changes can help the user to evaluate any subsequent effects on the quality of the data.

Table 3. Examples of challenges related to matrix effects, data quality, and logistics that can help determine the appropriate sensor selection.

[**Abbreviations**: DOC, dissolved organic carbon; mg N/L, milligrams N per liter; mg/L, milligrams per liter; mm, millimeter; n/a, not applicable; NTU, nephelometric turbidity units; NO_3, nitrate; UV, ultraviolet; <, less than; >, greater than]

Type	Typical values	Approach
		Matrix effects
High suspended sediment concentration/turbidity	>500 NTU	Use instruments with a shorter path length (that is, <10 mm) or deploy with a filtered flow path.
High DOC concentrations	>5–10 mg/L	Use instruments that measure the full UV spectrum.
	>30 mg/L	Use instruments that measure the full UV spectrum and use a shorter path length (that is, <10 mm).
High bromide concentrations	n/a	Use instruments that measure the full UV spectrum and include bromide compensation in algorithm.
High potential for biofouling	n/a	Use instruments with integrated or third party wipers.
		Data quality
High NO_3- concentrations	>20 mg N/L	Use instruments with a shorter path length (that is, <10 mm).
Low NO_3- detection limit needed	<0.5 mg N/L	Use instruments with a longer path length (that is, 10 mm or longer).
High NO_3- accuracy needed	<±0.5 mg N/L	Longer path length (that is, > 10 mm), full spectrum.
		Logistics
Buoy access only	n/a	Use instruments with integrated or third party wipers and can easily be integrated into existing data-collection platforms.
Infrequent site visits	<3–4 weeks	Use instruments with integrated or third party wipers.
Ease of use	n/a	Use instruments with integrated or third party wipers and "plug and play" controllers.

Instrument Performance Qualification

Prior to initial field deployment, the performance of UV nitrate sensors can be verified under controlled laboratory conditions. Visual and operational checks can also be performed after major instrument servicing or after shipping the instrument, because poor handling during shipping can disturb the alignment of optical components. Familiarity with the instrument specifications (table 2), the instrument manual, and this report can facilitate successful implementation of the performance qualifications.

First, there is a difference between the blank water that is appropriate for instrument blank analysis and reagent preparation compared to the blank water that is appropriate for instrument cleaning. As described in Chapter 2.0.3 of the USGS National Field Manual (Lane and others, 2003), inorganic-grade blank water (IBW) is suitable for use as blank samples to be analyzed for nutrients and other inorganic ions. The USGS National Field Supply Service (NFSS) sells IBW as stock number Q378FLD, and the USGS National Water Quality Lab (NWQL) provides results of acceptance testing for each lot sold *http://wwwnwql.cr.usgs.gov/qas.shtml?ibw*). This water, which is sometimes sold as "American Society for Testing and Materials (ASTM) Type 1 water," has very low concentrations of interfering species and nitrate plus nitrite (typically below 0.01 mg/L as N). If using laboratory blank-water systems, electrical resistivity must be greater than 18 mega-ohms per centimeter at 25°C (electrical conductivity less than 0.056 µS/cm at 25°C), and users can verify that the water is essentially free of inorganic constituents.

In contrast, distilled or deionized water (DIW) with electrical resistivity of at least 1.0 mega-ohms per centimeter at 25°C (electrical conductivity less than 1.0 µS/cm at 25°C) is specified by the National Field Manual (Wilde, 2004) for some equipment cleaning and other applications. If using multiple water types, clear labeling and field notes can help ensure that the appropriate water is used for the purpose. In addition, unopened bottles of blank water can be stored away from potential sources of contamination (vehicle exhaust, cleaning fluids, and other solvents) for later use.

The following inspection checks and test procedures make up the laboratory instrument performance qualification:

1. **Visual inspection**

    a. Ensure that all manufacturer's documentation, calibration reports, and serial numbers are recorded and maintained as described elsewhere (that is, National Field Manual).

    b. Visually inspect the instrument body for defects, blemishes, or imperfections and record all evidence of scratches, dents, nicks, or cracks.

    c. With the instrument off, visually examine the optical windows for scratching, pitting, staining, or misalignment with the sensor body.

    d. Inspect the electrical connector and associated cables for kinks, nicks, corrosion, or bent pins and contacts.

    e. Visually examine all ancillary components, such as wipers and controllers, for evidence of damage or corrosion.

    f. Record all inspection results (written and photographs) and contact the manufacturer immediately with concerns.

2. **Operational Inspection**

    a. Ensure that the most recent version of the operating software is installed and is operational.

    b. Following the manufacturer instructions in the user manual, apply power to the instrument and ancillary components to confirm that they are operational. Be aware that UV light can cause immediate and permanent eye injury, so it is important to never look directly into the measurement path when the instrument is operating.

    c. Verify and record the instrument reported nitrate concentration of the clean instrument in air and inorganic-grade blank water (IBW). Follow manufacturer recommendations to apply a baseline correction (that is, zero), if needed.

    d. Prepare or purchase a series of reagent-grade standard nitrate solutions for the range of concentrations expected at the deployment site. Verify that the accuracy is within acceptable limits and that the instrument response is linear across the range of nitrate concentrations. The accuracy can be calculated as the difference (absolute or percentage) between the measured concentrations and the known concentrations of the standard. Precision also can be verified against manufacturer stated values by calculating the standard deviation of the differences between the measured and known concentrations. Accuracy, precision, or non-linearity outside of the ranges specified by the manufacturer can be remedied by the user or the manufacturer. Additional information on calculating the accuracy, precision, and linearity (as well as detection limits) is in appendix 1.

    e. Repeat the accuracy, precision, and linearity checks with standard solutions spiked in natural waters (that is, matrix spikes) collected from the deployment site or a similar environment as described in appendix Water from the deployment site is likely to contain nitrate, so the acceptance criteria are based on the recovery of added nitrate rather than the measured value. Recovery of the nitrate spike is to be within accuracy specifications of the individual sensors in the absence of significant matrix interferences. If the matrix spike recovery is not within three times those specifications, further evaluation of the nature of the interferences (such as a dilution series) is warranted. Matrix spikes are to be used immediately upon preparation and cannot be kept for future reference because biological processes in the unfiltered sample can cause changes in the nitrate values rather quickly, especially at the lower end of the calibration curve. Also, evaluation of particle interference as a matrix effect requires careful suspension of sediment by using constant stirring and concurrent measurements of turbidity during the check.

f. If elevated turbidity or dissolved organic matter (DOM) is expected at the deployment site, but not present in the available matrix water, the performance of the instrument in the presence of these interferences can be verified by using standard reference materials. Although a variety of standard references are available, materials that are reflective of natural waters, such as Elliot Silt Loam soil or Suwanee River natural organic matter (available from the International Humic Substances Society, *www.humicsubstances.org*), are preferred to synthetic reference materials, such as formazin or polymer beads.

Instrument Deployment

A key consideration in the deployment of sensors is the identification of a stable, secure location representative of the water body of interest. Details on site selection and instrument deployment are provided in Wagner and others (2006), and those guidelines generally apply to the use of UV nitrate sensors. There are, however, some features unique to these instruments that warrant further attention.

Safety

Standard operating procedures for the safe deployment and operation of continuous monitors and water-quality samplers are always to be followed (for example, Lane and Fay, 1997). In addition, optical nitrate sensors use UV light, which can cause immediate injury and permanent damage to the eye. Therefore, it is important that users never look directly into the measurement path of these instruments while they are operating. Standard safety precautions, such as grounding rods and lightning suppressors, also are important to be used when deploying sensitive electronics at fixed sites.

Physical Infrastructure

As with other water-quality sensors, the current generation of UV nitrate sensors has been designed for either in situ deployment or a pumped configuration by using flow cells and sample collection vessels (fig. 13). General advantages and disadvantages of both in situ and pumped monitoring approaches are described in Wagner and others (2006). For UV-nitrate sensors, in particular, a major advantage of a pumped configuration is the ability to eliminate interference from particles by using a filter (typically 0.2 micron) in the sample flow path. In addition, instruments are typically accessible under all flow conditions, experience less biofouling, require shorter distances for power and communication, and allow for the possibility of sampling from multiple locations across the channel with one sensor. The primary disadvantages include costs and power requirements for pumps, frequent site visits to replace filters and tubing, and fouling within the tubing itself. Placing sensors on shore can increase or decrease sensor security, depending on the location and infrastructure present.

At some sites, an alternative could be the use of pipes or instrument cages that are deployed in the channel or on the channel banks (fig. 13). Pipes and cages can be constructed of stainless steel, aluminum, or other materials, depending on the corrosive nature of the site and protection needed from large debris. In addition, cages can be constructed to provide infrastructure for the simultaneous deployment of other sensors at the same location. The fixed infrastructure to which an instrument cage is mounted can be made of stainless steel or aluminum I-beams, channel or rails, with mounting brackets and lifting mechanisms of a variety of materials and designs. Advantages to this design generally include easy access for servicing, the ability to control the depth of instruments in the water column, space to deploy and control multiple instruments in close proximity, and the potential for adding ancillary equipment, such as pumps and filters, directly to the instrument cage. Primary disadvantages include the infrastructure and installation costs and concerns about the representativeness of measurements at a location typically close to the channel bank. In addition, pipes can create depositional environments that can affect the data quality or bury the sensors. Using buoys as "fixed" infrastructure is also an option, but boat access and solar power can be limitations to this approach in some systems.

Although the focus of this document is fixed monitoring sites, UV nitrate sensors have also proven valuable for longitudinal and depth profiling in rivers and lakes. These applications tend to be short-term studies that evaluate changes in water quality in relation to physical differences (such as temperature and density) that can affect chemical variability or that identify sources or sinks of nitrate. Profiling studies generally require that the sample output rate is well matched to the longitudinal or vertical variability in water quality. For example, vertical profiles of nitrate at a downcast rate of 1–10 cm per second require a rapid sampling rate (at least one sample every few seconds) to be effective at quantifying vertical variability. For a specific study site or application, users can check the depth limitations of their sensors and consider using sensors with deeper ratings if needed.

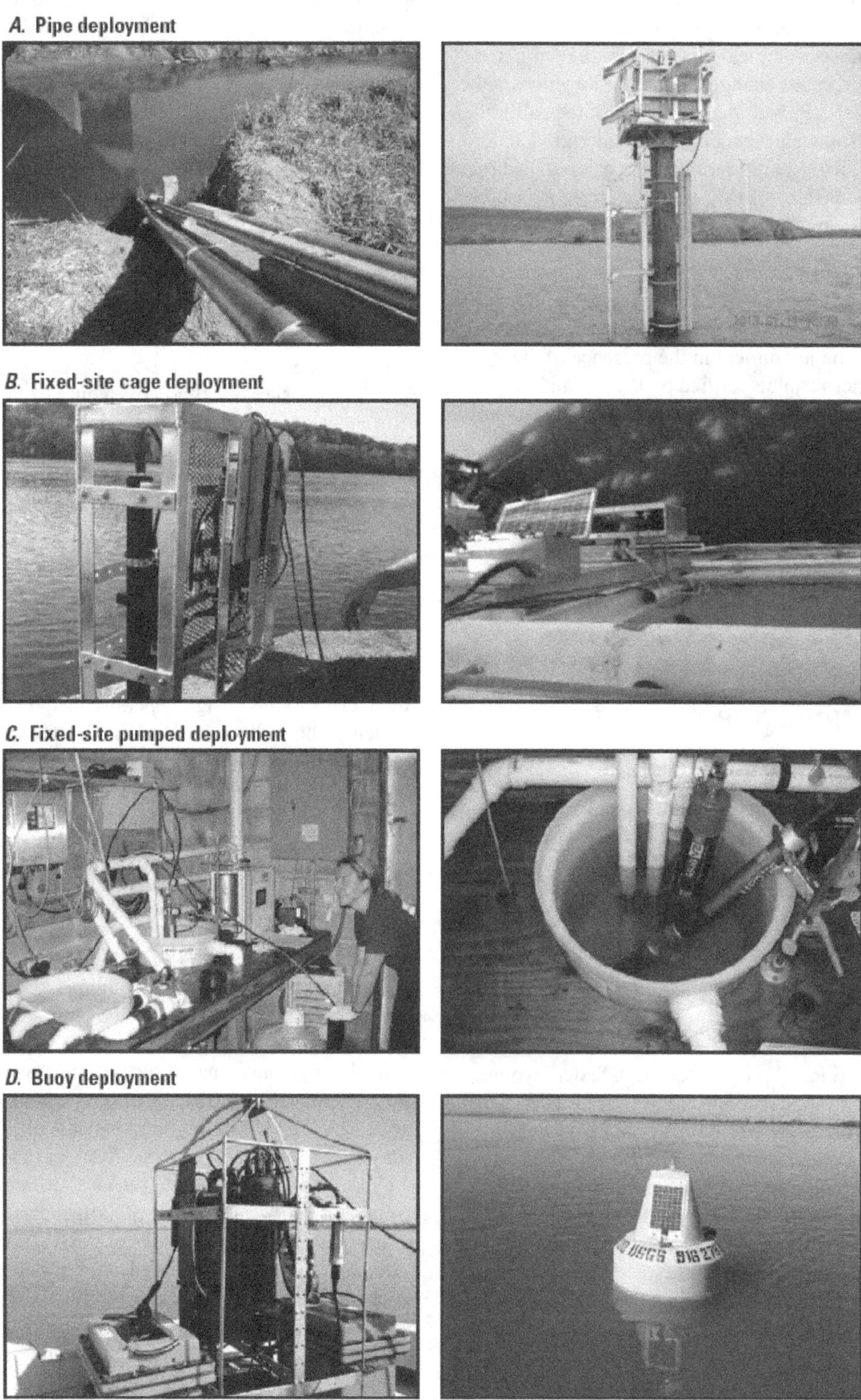

Figure 13. Ultraviolet (UV) nitrate sensor deployment infrastructure showing *A*, pipes; *B*, instrument cages; *C*, pumped applications (gage house); and *D*, buoy deployments. Photos courtesy of Jessica Garrett, U.S. Geological Survey; Jacob Gibs, U.S. Geological Survey; JohnFranco Saraceno, U.S. Geological Survey; Bryan Downing, U.S. Geological Survey.

Sensor Mounting

Manufacturers typically recommend mounting instruments horizontally in the water column with the gap open vertically in the sample flow path to decrease the accumulation of bubbles and sediments on the optical windows. Orientation is less of a concern, however, when using wipers or air blasts because sediment deposition and bubbles are typically dislodged prior to taking a measurement. Instruments can be mounted using corrosive resistant brackets that allow for a secure fit and limit torque and stress to the sensor housing or infrastructure. Hanging the instruments vertically without a specified mount is an alternative approach, but it can result in damage to the cables and connectors if not properly supported.

Anti-Fouling Measures

Anti-fouling measures come in two forms: passive chemical guards and mechanical devices (fig. 14). Copper is used in a broad range of in situ instruments because of its anti-fouling properties, and copper components and biofouling guards are available for some UV nitrate sensors. Other passive anti-fouling solutions, such as nano-polymer sprays, also can be useful if they do not interfere with the optical measurements. However, most manufacturers rely on mechanical devices (wipers or air blasts) to keep the optical flow path free of sediment, mineral precipitates, and biological growth. Wipers are made of nylon brushes or silicone blades that are integrated into the units or are external, add-on components that require separate power and control. Automatic air cleaning systems consist of compressors and tubing to deliver a short, concentrated burst of air across the windows in the optical chamber. Both wipers and air cleaning systems consist of field serviceable components that need to be inspected during each site visit and cleaned or replaced per manufacturer recommendations.

For instruments deployed in a gage house or other structure, biofouling is generally managed through the intermittent pumping of samples through a chamber or flow cell. However, an inorganic film has been observed on the optical windows when the total dissolved solids were greater than approximately 150 mg/L (Jack Gibs, U.S. Geological Survey, written commun., 2012). Users, therefore, may consider use of an automated blank water spray or other anti-fouling techniques to prevent changes to the optical properties of the windows during intermittent wetting and drying cycles.

Dataloggers and Controllers

The current generation of UV nitrate sensors uses a combination of integrated microcontrollers that have designated functions ranging from lamp excitation, photo detection, and pulse and waveshaping of the raw signal to producing digital or analog output and controlling internal or external anti-fouling functions. The output of certain sensors utilizes conventional, industry standard transmission protocols (that is, analog, RS-232, RS-485, or SDI-12) allowing the user to select from a variety of field DCPs. In most installations, digital output is preferable to analog output because analog data cannot be routed over long (that is, more than 100 ft) cable distances, and analog data have to be digitized or level shifted before they can be telemetered.

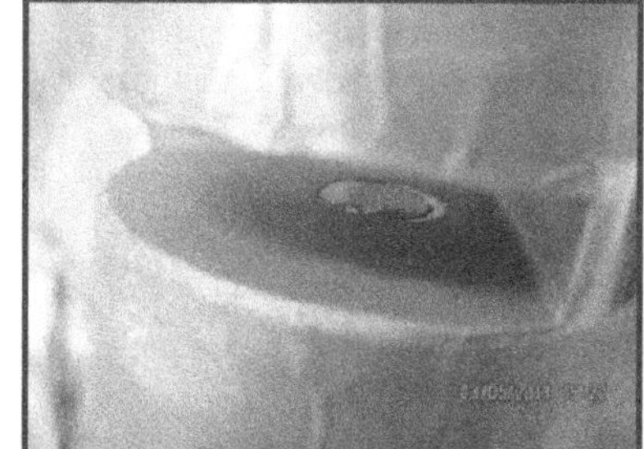

Figure 14. Examples of passive (copper guard, copper tape) and active (mechanical wiper, air blast system) mechanisms to reduce the effects of biofouling on ultraviolet (UV) nitrate sensors. Photos courtesy of Richard Cartwright, U.S. Geological Survey; Joseph Bell, U.S. Geological Survey, and Justin Irving (s::can).

Several UV nitrate sensors utilize proprietary transmission protocols that also require the use of additional external controllers. Because these controllers typically are not submersible, a location (usually within 100 feet) is needed for mounting and powering the units. The use of submersible DCPs capable of controlling sensors and logging data is a good option in some settings, and it results in fewer cables and greater flexibility for the simultaneous deployment of a variety of sensors.

Power and Communication

Electrical connections are a vital part of all instrumentation systems, playing the critical role of conducting uninterrupted electrical power to the instruments and for communication with the instruments. As connectors and cables are submerged in natural waters, exposed to harmful effects of UV light, and subject to oxidation, they can rapidly degrade and affect the transmission of data. As nitrate sensors generally require ample stable current at the manufacturer specified voltage (table 2), the length of the power cable is of prime importance. Approaches to correct for voltage drops and power loss over long cable distances range from adjustable power supplies or convertors designed to maintain a steady voltage at a flexible voltage input range. Most UV nitrate sensors have electronic circuitry to protect against low-power conditions or when suboptimal input voltage or current is detected. The flexible power supply could be installed at either end of the power cable, as long as the instrument is provided the correct voltage.

As a rule of thumb, cables either can be purchased from or are recommended by the manufacturer. Cable insulation is an important consideration because the cable is subject to the rigors of flowing waters and, in some cases, is used for lifting purposes. Many nitrate sensor manufacturers supply neoprene jacketed cable, which commonly is used in temporary underwater deployments and can be rapidly damaged in high-energy environments. A cable made with an outer jacket of polyether-polyurethane can be used in such environments. If the sensor deployment requires custom cables, the type and design is to be considered carefully. For example, if data from the sensor are analog output, then ultralow resistivity and capacitance cable is specified. Further, if the cable will be used as a lifting mechanism for the sensor (not recommended), a reinforced cable with a core of Kevlar or similar material can be used.

Data Collection

The collection of data from UV nitrate sensors has two relevant time intervals: (1) the sampling interval, and (2) the reporting interval. The sampling interval describes the number of samples collected over a defined time span, which is typically used to calculate a single mean or median concentration for reporting. Sampling intervals range from 1 to 120 seconds for the current generation of UV nitrate sensors (table 3) and can be reported in frequency units (hertz, Hz) that describe the number of cycles per second. The collection of instrument data at a high rate (referred to as "burst sampling") can be used to filter instrument noise and account for the chemical variability in water passing by the sensor during the measurement period. In addition to calculating a mean or median value for reporting, burst sampling can also be used to calculate statistics that describe the uncertainty of the measured value (such as the relative standard deviation) and help identify instrument performance issues or matrix effects when statistical thresholds are exceeded. For example, an increasing trend in the relative standard deviation of the burst data over time can be indicative of lamp degradation, fouling of the optical windows, or high particle concentrations (fig. 15).

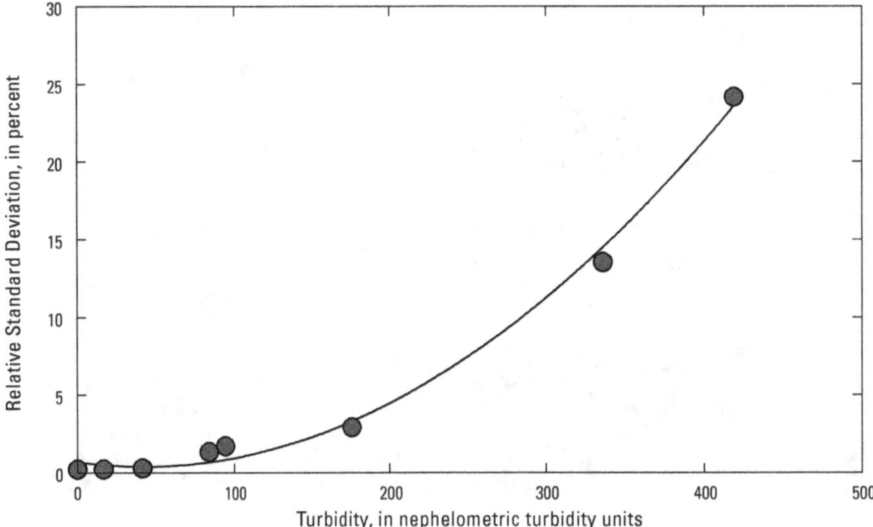

Figure 15. Example of the effect of increasing turbidity on the relative standard deviation of nitrate concentrations collected during a laboratory test of a 1 milligram per liter as nitrogen matrix spike solution. Data were collected with a 10-millimeter Satlantic SUNA sensor and averaged over 2 minutes with a sampling interval of about 2 seconds.

The reporting interval describes the time between reported values. Reporting intervals for USGS data are typically between 15 and 60 minutes for discharge and continuous water-quality parameters such as pH and turbidity. In many cases, the reporting value represents a single instantaneous measurement taken at the specified reporting interval, in which case the sampling interval is the same as the reporting interval. Although there is no specific guideline for the appropriate reporting interval, a general rule for UV nitrate sensors is to report data at time intervals that are frequent enough to capture the rate of change or variability in nitrate concentrations, but not so frequent as to result in excessive power consumption or rapid instrument-lamp degradation. The user can determine the optimum reporting interval at each site on the basis of the variability in discrete and continuous data and the overall study goals. For example, the optimum reporting interval for calculating monthly or annual loads would not be the same for studies of nitrate cycling as for studies of storm event dynamics.

Maintenance and Field Operations

The maintenance and operation of UV nitrate sensors is to follow manufacturer recommendations and existing USGS protocols for continuous monitors (for example, Wagner and others, 2006). The general operations include instrument maintenance, sensor inspection and calibration checks, field cleaning, and troubleshooting. Other issues related to sensor operation, such as site selection and cross-section surveys, are covered in Wagner and others (2006) and are critical to the collection of representative nitrate concentrations with UV sensors.

Maintenance

Maintenance includes a variety of functions that can be performed both remotely, for sites with telemetry, and during field visits (table 4). The daily review of data and performance diagnostics for both the sensor and the system is easily completed remotely, considering that most sites are equipped with telemetry, satellite data transmission capabilities, or both. The use of satellite transmission alone can put limits on daily maintenance because of the amount of data that can be transmitted in a short period (approximately 10 seconds per site) and a lack of two-way communication with the data collection platform (DCP) that prevents making queries or changes remotely. Nevertheless, satellite telemetry can be less susceptible to the communication problems inherent to telemetered systems at some sites and can be important when real-time data delivery is critical to project objectives.

Field maintenance for continuous monitors generally includes a site inspection, instrument inspection and cleaning, field-blank or calibration checks, and discrete sample collection. Although the USGS does not provide specific instructions on the frequency of field visits, generally site visits are made as often as needed to ensure that data-quality objectives are met. In practice, the frequency of field visits to maintain continuous monitors is often a combination of fixed and data-dependent maintenance schedules. Fixed schedules are typically determined by discrete sample collection schedules rather than by routine sensors maintenance. Data-dependent schedules can be more useful in sites where fouling rates are not known (or vary significantly) or where there is periodic disruption from sensor malfunction, sedimentation, pump failures, debris, ice, or vandalism (Wagner and others, 2006).

Although the optimal schedule for site visits to maintain UV nitrate sensors will vary by site, season, and deployment approach, the use of wipers and air cleaning systems will extend instrument maintenance intervals. Still, a maximum period between servicing (typically less than 4–6 weeks, but up to 12 weeks in cold or very low-fouling environments) can be implemented for site inspections and discrete sample collection. However, sites with very high biological productivity or without remote data access could require more frequent service (for example, every 2–4 weeks), whereas sites with data-quality objectives that require a greater degree of accuracy could require maintenance of water-quality monitors at even shorter intervals.

Table 4. General maintenance operations at an ultraviolet nitrate-sensor deployment site (modified from Wagner and others, 2006).

Daily maintenance operations (for sites with remote access)
Daily review of nitrate data, ancillary data, and data-quality indicators (if recorded).
Daily check of sensor performance (lamp hours, instrument noise, instrument temperature).
Daily check of system performance (power, datalogger storage, data transmission).
Flagging of spurious data, if necessary.
Modifications or queries to the data-collection platform, if needed.
Maintenance operations during field visits
Inspection of the site for signs of physical disruption.
Inspection and cleaning of sensors for fouling, corrosion, and damage.
Battery (or power) check.
Time check.
Routine sensor cleaning and servicing.
Sensor baseline check (blank correction, if necessary).
Downloading of data (if necessary).
Sensor linearity check (optional).
Discrete sample collection (cross-section or near sensor).

Field Protocols

Field protocols ensure that sensors are working properly and provide critical data for interpreting and processing the water-quality record. They include visual inspections, instrument cleaning, calibration checks, data downloads (if appropriate), and documentation. The standard protocols defined by Wagner and others (2006) for continuous monitors apply to the field servicing of UV nitrate sensors and are to be followed when possible and documented when not possible. There are, however, several important differences between UV nitrate sensors and many of the traditional continuous water-quality probes, such as pH and temperature, that affect field protocols. For example, UV photometers have a sensitive, high-power lamp that degrades over time and can be compensated for directly by the instrument (for instruments with a reference channel) or by the user. In addition, the desired output (nitrate concentration) is derived from an algorithm that can be influenced by a variety of dissolved inorganic and organic substances and particles in the sample path. Finally, the new generation of sensors provides opportunities to collect information about instrument performance that can ultimately improve data quality through early identification of problems and a subsequent reduction in data gaps.

Although field servicing and maintenance are necessary, users can determine which steps need to be completed under field conditions and which steps are better accomplished in a lab setting. For example, blank-water calibrations to account for instrument drift can be better completed in a controlled lab setting rather than under potentially harsh field conditions (that is, rain, snow, wind, or dust). Despite the cost of purchasing a replacement sensor, replacing instruments can also be a useful alternative when field servicing procedures are not possible or when more detailed troubleshooting is needed. However, to obtain comparable data, it is important to ensure that the sensors are comparable in terms of settings and design, and that the sensor to be deployed has been characterized as described in the "Instrument Performance Qualification" section of this report.

The following protocol is for the general field servicing of UV nitrate sensors. Users can also carefully evaluate manufacturer recommendations for servicing because each instrument differs in terms of design and data specifications. Field servicing of UV nitrate sensors will often require a portable power supply, a controller (laptop or other device), and the availability of shade to avoid extreme heating of the sensor while it is out of the water. In addition, servicing instruments deployed in a gage house requires careful inspection of all system components (pumps, tubing, flow cells, and optical windows) to ensure that the data meet quality objectives (L.S. Feinson and others, U.S. Geological Survey, written commun., 2011).

1. **Perform a site inspection**. Observe and document in field notes (hand-written or electronic) and photographs of the site conditions and state of the infrastructure. Document any changes in the biological, physical, and hydrologic conditions at the site, such as high algal productivity, exposed channel, stagnant water, increased light exposure, or any signs of vandalism. Note the condition of the sensor infrastructure in the water such as damaged cages or pipes, corrosion, snagged debris, or changes in instrument depth since last deployment.

2. **Collect a field meter reading (if available)**. An independent field meter serves as a check on the deployed sensor and can document changing environmental conditions during servicing. A field meter can also be used for cross-section surveys or other measurements that verify the data collected by the fixed-site sensor are representative. If a second UV nitrate sensor of the same design is available, measurements can be made next to the fixed sensor for several sampling intervals, following guidelines in Wagner and others (2006). If a field meter is not available, an alternative approach is to collect a discrete sample concurrent with and next to the fixed sensor for laboratory analysis by using a separate UV nitrate sensor or traditional wet-chemical methods.

3. **Collect a field measurement with the fixed sensor**. Recording a field measurement with the fixed sensor provides an initial measurement that can be compared to field measurements after cleaning to determine the correction needed for fouling (step 7). Time, field readings, and monitor condition can be recorded by using standard USGS field forms and protocols. At sites with rapidly changing environmental conditions—for example, a change that exceeds the calibration criteria (table 5; Wagner and others, 2006) within 5 minutes—ambient water can be collected in a clean 5-gallon dark bucket (or similar) and used as a stable environment for readings before and after cleaning. Because settling particles within the bucket could introduce some uncertainty into the measurements, filtered water can be used in this step if there are high suspended-sediment concentrations. If this is the case, a small submersible pump, high-quality tubing, and a high-volume capsule filter (pore size less than 45 micron) can be used for sampling.

4. **Remove sensor from the monitoring location**. Ensure that the method for removing the instrument from the water is appropriate for the sensor design and does not cause damage to the cables, connectors, or the sensor itself. Lifting the sensor by the electronics cable can cause damage to the cable.

5. **Perform a sensor inspection**. Observe and document in field notes and photographs the condition of the instrument, cables, connectors, wipers, and infrastructure. This includes a careful inspection of the nitrate sensor for (1) fouling, staining, or scratching on the optical windows; (2) degraded seals around the optical windows or wiper housings; (3) damage to the sensor housing and connectors; (4) damage or wear of the wiper assembly,

Table 5. Calibration criteria for inorganic blank water checks of the ultraviolet (UV) nitrate sensors based on manufacturer stated accuracy in a zero milligram per liter (mg/L) as nitrogen (N) solution for different sensor models (Hach) or wavelengths (Satlantic, s::can, TriOS).

[Only the wavelengths commonly used for natural waters are shown for the s::can. **Abbreviations**: mg/L as N, milligram per liter as nitrogen; mm, millimeter; %, percent]

Pathlengths	Stated accuracy	Calibration criteria in inorganic-free blank water (mg/L as N)
	Hach Nitratax	
1, 2, 5	±3-5% of reading or 0.5–1.0 mg/L, whichever is greater	–0.5 to +0.5 (plus, clear) –1.0 to +1.0 (eco)
	Satlantic SUNA	
5, 10	±10% of reading or 0.03–0.06 mg/L, whichever is greater	–0.03 to +0.03 (10 mm) –0.06 to +0.06 (5 mm)
	S::CAN spectrolyzer	
0.5–100	±2% of reading plus 1/optical path length (in mm; mg/L)	–0.03 to +0.03 (35 mm) –0.2 to +0.2 (5 mm) –0.5 to +0.5 (2 mm)
	TriOS ProPS	
1–60	±2% of reading or ±0.155 mg/L, whichever is greater	–0.155 to +0.155

brushes, or other anti-fouling components; and (5) cracks, cuts, kinks, or wear on the cables. The infrastructure also can be more closely inspected for signs of damage or extensive corrosion. Users are to observe that mechanical parts, such as wipers or pumps, are functioning properly. If the instrument is being used in a flow-through design, additional inspection of the pumps, tubing, filters and filter housings, flow cells and collection vessels can be performed to detect damage, clogging, or fouling. Additional tasks can include tightening screws, checking internal moisture readings, waterproofing connections, and replacing serviceable parts, such as wipers (according to manufacturer specifications). Some routine maintenance is not be possible in the field, requiring temporary removal of equipment.

6. **Clean sensors.** Field cleaning of optical sensors requires extra diligence with regard to the optical windows and the measurement path. Abrasive cleaners are never to be used, and the optical windows may only be cleaned with lint-free lens paper or soft-bristled brushes. With appropriate safety precautions and waste collection, a weak hydrochloric acid solution (less than 5 percent) or ethanol can be used in the field to efficiently remove staining or severe fouling, including the iron and manganese precipitates that form in some systems. In addition, the use of chemical cleaners and detergents is always followed by a distilled or deionized water (DIW) rinse. DIW has an electrical resistivity of at least 0 mega-ohms per centimeter at 25°C (electrical conductivity less than 0 µS/cm at 25°C) and is specified by the National Field Manual (Wilde, 2004) for some equipment cleaning and other applications. The sensor housing, cables, and wipers can also be cleaned with DIW, and pump tubing (if used) that shows signs of fouling can be replaced.

7. **Return the sensor to the monitoring location and perform fouling check.** Return the sensor to the water (in rapidly changing environments, ambient water in the bucket) and measure the nitrate concentration to make the necessary fouling corrections, as described in Wagner and others (2006).

8. **Remove sensor and perform a baseline calibration check.** Field baseline checks with inorganic blank water are used to monitor for sensor integrity and correct for baseline drift primarily due to the degradation of the UV lamps over time. Although not required during every site visit, performing a baseline calibration check at least once every 4–6 weeks can help ensure data quality. After a thorough DIW rinse of the optical path, follow manufacturer specifications to insert inorganic-grade blank water (IBW) into the measurement path and measure the nitrate concentration. As noted previously, the USGS National Field Supply Service (NFSS) sells IBW as stock number Q378FLD, and the USGS National Water Quality Lab (NWQL) provides results of acceptance testing for each lot sold *http://wwwnwql.cr.usgs.gov/qas.shtml?ibw*). Take care to avoid introducing bubbles into the flow path, which can result in erroneous readings.

Baseline calibration checks that show nitrate concentrations within the manufacturer's accuracy specifications (for example, "calibration criteria," table 5) indicate that significant drift has not occurred and the instrument does not need a baseline calibration. Values outside of this range require a baseline correction (for example, a new blank spectra file or blank measurement) that can be entered according to manufacturer specifications. Prior to making this correction, however, it is critical to be sure that the instrument is clean and the measurement is not affected by bubbles or direct sunlight. Therefore, users are to examine and clean the optical flow path and repeat the blank check with a separate aliquot of IBW from a previously unopened bottle to confirm that the measured value is still outside of the acceptable range prior to applying a baseline correction. If a baseline correction is entered, confirm that the new blank value falls within the specified calibration criteria.

For UV nitrate sensors with a reference channel (table 2), baseline corrections for lamp degradation usually are not necessary because the reference beam is used to compensate for drift with time. Sensors without a reference channel, but in good working order and running intermittently, could need to be recalibrated only once every few months. The frequent need for recalibration or high exceedances of the calibration criteria (for example, greater than three times the expected blank range) can be investigated further because this could indicate a problem with the sensor, blank water, or servicing procedures.

9. **Linearity checks with nitrate standards (optional).** The linearity of the instrument response at a range of nitrate concentrations can be verified in the lab prior to use (see "Instrument Performance Qualification" section), but additional field linearity checks with nitrate standards typically are not necessary. However, if a quality-assurance plan requires additional nitrate standard checks in the field to verify the linearity of the instrument response, it is important to note the following:

    - Nitrate check standards are to be within the range of concentrations typically measured for a given site.
    - Nitrate check standards are to be made with American Chemical Society (ACS) reagent grade chemicals spiked into inorganic-grade blank water within 48 hours of use and stored in a dark, cold (4°C) location or purchased as certified solutions in sealed containers to avoid biological growth and other sources of contamination.
    - Precautions are to be taken when using potassium nitrate (KNO_3) or other nitrate standards in the field, including the use of protective equipment (safety glasses and latex gloves) and waste disposal as described in the materials safety data sheet (MSDS).

If the measured nitrate concentration in the linearity check standard is not within the stated instrument accuracy (table 2), inspect the instrument optical path for bubbles or other sources of contamination and repeat with the nitrate standard. If the instrument check is still out of specifications, return the sensor to the laboratory or the manufacturer for further evaluation. Users can consider performing periodic checks with nitrate standard added to matrix waters, which will be particularly important in systems where matrix effects are expected to vary significantly with time. However, the inherent difficulty in performing a controlled matrix spike under field conditions—particularly to evaluate the effects of suspended particles—could preclude this test from being performed in the field by most users. Instead, matrix effects can be evaluated under laboratory conditions, as described in the "Instrument Performance Qualification" section and appendix 1 of this report.

Troubleshooting

As with all continuous monitors, there are a range of possible issues with the electronics or ancillary components that can affect UV nitrate sensor performance in the field. Troubleshooting issues in the field reduces sensor down time, the need for data corrections, and additional trips to the site; however, more difficult problems that require sensor testing or technical support can be better addressed in the laboratory. Some of the more common problems with UV nitrate sensors are included in table 6, but users can follow manufacturer recommendations for troubleshooting. It is also important to note that UV nitrate sensors and controllers can store or output a variety of data that can prove useful for diagnosing and troubleshooting sensor performance issues. For example, information can include lamp run time, the temperature of internal components, and statistics on the algorithm fit (table 7). Diagnostic parameters vary by manufacturer and are typically described in instrument user manuals.

Data-Processing Procedures

Data-processing procedures for continuous nitrate data follow existing USGS methods for continuous monitors described in Wagner and others (2006) and other related guidance. Continuous records undergo an initial evaluation, flagging of erroneous data, application of data corrections, and a final evaluation. Data are archived electronically in the National Water Information System (NWIS) database with project or site information following project and Science Center data-management and quality-assurance plans. New data received by way of telemetry or otherwise recently loaded into NWIS are viewed each day, and any obviously erroneous data are addressed. Data corrections are applied for fouling and drift on the basis of information collected during site visits. In addition, corrections for systematic errors due to matrix effects can also be applied. A final data evaluation includes a review of the record, review of the corrections, and any final revisions.

Uncorrected data from the optical sensors are valuable information that can be used for data-processing or for additional research or interpretation. Any raw data used for data-processing procedures are archived primarily in NWIS or in Science Center records, if NWIS is unable to store the necessary information. For UV nitrate sensors, this can include burst data, raw absorbance values at individual wavelengths (if reported), and instrument diagnostic data.

Fouling and Drift Corrections

Fouling and drift corrections for continuous water-quality sensors are needed when the sum of the absolute errors exceeds specified criteria for water-quality data corrections (Wagner and others, 2006). Corrections can also be applied

Table 6. Common issues and guidelines on troubleshooting for ultraviolet nitrate sensors.

[**Abbreviations**: mg/L as N, milligram per liter as nitrogen; mg/L as NO_3, milligram per liter as nitrate; NA, not applicable]

Questionable data characteristic	Possible cause if valid data	Possible cause if erroneous data	Troubleshooting test or possible solution
Intermittent spikes.	Point source or other brief episodic inflow of nitrate.	Interferences or fouling.	Verify sensor flow path is clear. Clean sensor windows. Check anti-fouling system performance.
Diurnal nitrate variability.	Nutrient cycling.	Stray light interference. Solar induced power fluctuations.	Adjust sensor position. Collect independent samples throughout daily cycle.
Data do not track with co-located continuous sensors.	Expected relation between parameters does not apply for the site or time period.	Multiple sensors not measuring the same representative water. Sensor fouling.	Clean and calibrate all sensors. Collect cross section water quality data. Collect discrete samples at the sensor location.
Data are negative values.	Nitrate concentrations below the detection limit.	Incorrect offset/blank sample. Salinity correction misapplied.	Clean and evaluate blank spectra (upload new spectra if needed). Check for salinity compensation. Collect independent samples.
Sensor errors or warnings.	Non-critical errors.	Critical errors influencing data quality.	See manufacturer user manual for specific error and remedy.
Standard solution value is off by a factor of 4.5.	NA.	Sensor reporting in incorrect units (mg/L as N and mg/L as NO_3).	Correct reporting units if necessary. Analyze a new standard solution.
Standards not within instrument specifications.	Bad standards. Standards outside of instrument range.	Instrument failure.	Return to lab for evaluation with new standards. Manufacturer calibration may be needed.
Sensor and transmitted values not in agreement.	Incorrect clock settings.	Faulty wiring or programming problem.	Check sensor output settings. Check connections and equipment/DCP clock settings.
Noisy data.	Variability in nitrate over short-time periods.	Lamp degradation. Intermittent fouling. Matrix interferences present.	Check lamp hours and indicators. Clean and evaluate sensor blanks. Collect a lab sample to evaluate interferences.
Significant drift.	None.	Lamp degradation.	Apply drift corrections if within calibration criteria. Return to manufacturer for lamp replacement.
No signal.	Good data could be present on sensor if internal logger is present.	Bad cables, connectors or lamps. Inadequate power to sensors.	Check voltage supply to sensor. Check cables and connectors for damage. Confirm sensor response using host software.

Table 7. Examples of ancillary information useful for troubleshooting ultraviolet nitrate sensor performance.

Parameter	Description
Temperature	Temperature of the instrument components (lamp housing, spectrometer).
Power	Voltage of the lamp power supply, externally applied voltage, internal regulator.
Humidity	Humidity inside the instrument housing.
Fitting parameter	Diagnostic of the instrument algorithm fit.
Lamp time	Cumulative time of the lamp on.
Error messages	Indicators of instrument performance.
Spectral values	Intensity of the light source.
Dark value	Spectral value during dark measurements.
Absorbance	Absolute absorbance values in the ultraviolet and visible range.
Instrument settings	Instrument set up.

and entered into NWIS for a lower criteria or threshold at the discretion of the scientist. Data-correction criteria in this report for UV nitrate sensors are based on the uncertainty of the manufacturer-stated accuracy at a given concentration (table 5). For example, data correction is recommended with the SUNA sensor if the sum of the absolute errors due to fouling and calibration drift (that is, total error) is greater than 0.03 mg/L as N or 10 percent of the measured concentration, whichever is greater. However, these criteria are minimum requirements for data corrections; users can evaluate instrument accuracy in the lab to determine if more stringent criteria could justify non-zero data corrections.

Fouling corrections can be needed to account for reductions in transmittance due to biological growth and sediment deposition on the optical windows of UV nitrate sensors. Corrections are most critical when anti-fouling measures, such as wipers or air blasts, are not used. When they are used, fouling corrections are likely to be limited to periods with wiper failures or extreme biological fouling that is not easily removed by mechanical action.

Determination of the error due to fouling follows Wagner and others (2006):

$$E_f = (M_a \quad M_b) \quad (F_e \quad F_s) \tag{4}$$

where

- E_f is the fouling error,
- M_a is the fixed-sensor nitrate concentration after cleaning,
- M_b is the fixed-sensor nitrate concentration before cleaning,
- F_e is the field-meter nitrate concentration at the end of fixed-sensor servicing, and
- F_s is the field-meter nitrate concentration at the start of fixed-sensor servicing.

As noted previously, the availability of a separate field meter can be limited because of instrument costs. Therefore, if using a bucket or taking field readings where conditions are not rapidly changing, $F_e - F_s$ is about zero.

Corrections for UV nitrate sensor drift can be needed periodically to account for the loss of light intensity as the lamps degrade. Instruments with a reference or compensation beam (table 2) automatically compensate for drift due to aging of the lamp or other electronic disturbances and, in theory, do not need calibration for instrument drift. However, all continuous water-quality sensors can be evaluated for drift according to this calculation from Wagner and others (2006):

$$E_d = V_s \quad V_c \tag{5}$$

where

- E_d is the calibration drift error,
- V_s is the known concentration of a standard or solution containing nitrate, and
- V_c is the nitrate concentration reported by the sensor in the standard or solution.

For UV nitrate sensors, instrument drift can be determined in inorganic blank water (IBW) as described in step 8 of the "Field Protocols." The calibration criteria in table 5 are used to determine whether a baseline correction is needed, which is achieved by either uploading new reference spectra or resetting the blank value using the instrument software, according to the user manual.

If the fouling error exceeds the calibration criteria (table 5), a two-point variable data correction for nitrate can be used to linearly interpolate the necessary correction on the basis of the percentage error for the range of recorded values. This is useful because a one-point data correction can give unreasonable (negative) results, particularly at low nitrate concentrations. The two-point variable data correction when $F_e - F_s \sim 0$ is calculated as follows:

$$\%C_f = 100\left(\frac{(M_a \quad M_b)}{M_b}\right) \tag{6}$$

where

- $\%C_f$ is the fouling correction (in percent),
- M_a is the fixed sensor nitrate concentration after cleaning, and
- M_b is the fixed sensor nitrate concentration before cleaning.

If a calibration is required, and the field-meter readings vary during servicing (that is, $F_e - F_s$ is a non-zero value), the correction is calculated as follows:

$$\%C_f = 100\left(\frac{(M_a \quad M_b) \quad (F_e \quad F_s)}{M_b}\right) \tag{7}$$

where

- F_e is the field-meter nitrate concentration at the end of fixed-sensor servicing, and
- F_s is the field-meter nitrate concentration at the start of fixed-sensor servicing.

In addition, a correction—zero or non-zero—is typically required in NWIS for every site visit. Corrections can be made at values less than the calibration criteria at the discretion of the scientist.

A common assumption for fouling corrections is that equipment fouls at a constant rate that begins immediately after the last cleaning and, therefore, represents the starting point for a correction. This assumption is not always valid, particularly when there are wiper malfunctions or episodes of high biological productivity during deployment. It also can be possible to identify the actual start of fouling by using sensor diagnostic data or data statistics. For example, gradual or abrupt changes in the standard deviation or lamp intensity can be indicative of fouling events. If the start of a fouling event is identified, corrections can be applied from that date.

Bias Corrections

The UV absorption method for nitrate is a direct spectrophotometric measurement, and UV nitrate sensor concentrations, therefore, would be strongly correlated with laboratory nitrate measurements (within instrument specifications) in the absence of interferences. However, the presence of dissolved organic matter, bromide, or suspended particles in the matrix water can result in systematic errors, or bias, in relation to laboratory nitrate measurements. This can be especially problematic if sensors are calibrated to nitrate standards without accounting for interferences in natural waters (Drolc and Vrtovšek, 2010). These errors—which are related to instrument design and typically result in a positive bias of the UV sensor nitrate concentrations—can be accounted for through the careful application of corrections to the sensor data.

Two general approaches can be used to correct for a systematic error (bias) with the UV nitrate sensor: (1) data corrections based on continuous in situ measurements of interfering substances, or (2) corrections based on the correlation of continuous measurements with nitrate concentrations from discrete water-quality samples measured in the laboratory. The first approach uses the data collected during laboratory matrix spikes (described in "Instrument Performance Qualifications") to evaluate interferences and develop bias corrections. For example, continuous measurements of chromophoric dissolved organic matter fluorescence (FDOM) and turbidity can be used to account for organic matter and suspended particles, respectively, in instruments susceptible to these types of interference. This approach allows for a clearer mechanistic understanding of matrix effects and the potential for real-time corrections, but it requires a careful characterization of nitrate sensors across the full range of matrix conditions expected for the site. In addition, this approach requires the characterization, deployment, and maintenance of additional in situ sensors and quality assurance of the data to apply accurate corrections.

The second approach—often referred to as a "local calibration"—relies on a comparison of in situ and discrete nitrate concentrations measured at the same time to quantify the bias in the sensor data. This approach has several advantages compared to a full instrument characterization (easy to apply, no sensor testing, no additional instrument deployments, and so on), but several key assumptions need to be met:

1. The error is systematic (for example, sensor concentrations are always biased in one direction, typically high) rather than random.
2. The sample size is sufficiently large (greater than 20 samples) to allow for comparisons.
3. The relationship between the in situ and discrete concentrations has a high coefficient of determination (R^2 value greater than 0.8).
4. The slope of the regression is close to one (0.9–1.0), or variation across the range of nitrate concentrations can be easily accounted for.

Hypothetical examples of data sets that would violate the stated assumption are shown in figure 16. In some situations where one or more assumption is violated, bias corrections can still be considered with proper justification. For example, a study with n less than 20 can still consider applying a bias correction if all other assumptions are met. Some data sets also require more rigorous statistical analyses for slope comparisons and calculating bias. However, if these assumptions are met, the need for a correction can be determined based on the mean bias:

$$B_c = \frac{1}{n}\sum_{i}^{n}(S_i \quad D_i) \tag{8}$$

where

B_c is the mean bias in concentration units,
S_i is the sensor nitrate concentration in the i-th sample,
D_i is the discrete nitrate concentration in the i-th sample, and
n is the number of sensor values.

For practical purposes, S_i is the fouling- and drift-corrected UV nitrate sensor concentrations, whereas D_i is the laboratory nitrate plus nitrite concentrations (that is, USGS parameter code 00631). Although the current generation of sensors does not explicitly measure nitrite, the nitrate calculation is influenced by cross-sensitivity to nitrite, and, therefore, it is to be included.

If assumptions are met, the mean bias (B_c) calculated previously can be compared to the manufacturer (table 8) or user-specified calibration criteria (that is, within the accuracy calculated as in appendix 1) to determine if a bias correction is warranted. The following are general guidelines:

- If the mean bias (B_c) is within the manufacturer or user-specified calibration criteria, a correction is not needed.
- If the mean bias (B_c) is one to three times the manufacturer or user-specified calibration criteria, then a calibration to discrete data is acceptable.
- If the mean bias (B_c) is more than three times the calibration criteria, a calibration is generally not appropriate. Instead, the instrument can be returned to the laboratory or manufacturer for further evaluation of matrix effects and sensor performance.

Applying a bias correction on the basis of instrument accuracy, instead of the magnitude of the mean bias, implies that instruments with better accuracy are more susceptible to interference. For example, the maximum error for bias correction (table 8) for a 1.0 mg/L as N solution would range from 0.18 to 3.0 mg/L as N for the different instruments and path lengths available. Therefore, caution is warranted when comparing the relative sensitivity of instruments to matrix interferences.

Uncertainty in the laboratory nitrate analysis can influence the correlation between in situ nitrate-sensor data and discrete-sample data. Although the relative standard deviation of repeated measurement is typically less than 3 percent according to current laboratory instrumentation and methods (Patton and Kryskalla, 2011), users could include laboratory uncertainty when evaluating the need for bias corrections. In addition, calibrations with cross-section average nitrate concentrations are potentially not possible in poorly mixed systems. In such cases, additional discrete sampling next to the sensor could be used for bias corrections, although site selection can also be re-evaluated. These decisions are ultimately left to the best professional judgment of the scientist and are to be documented according to USGS protocol.

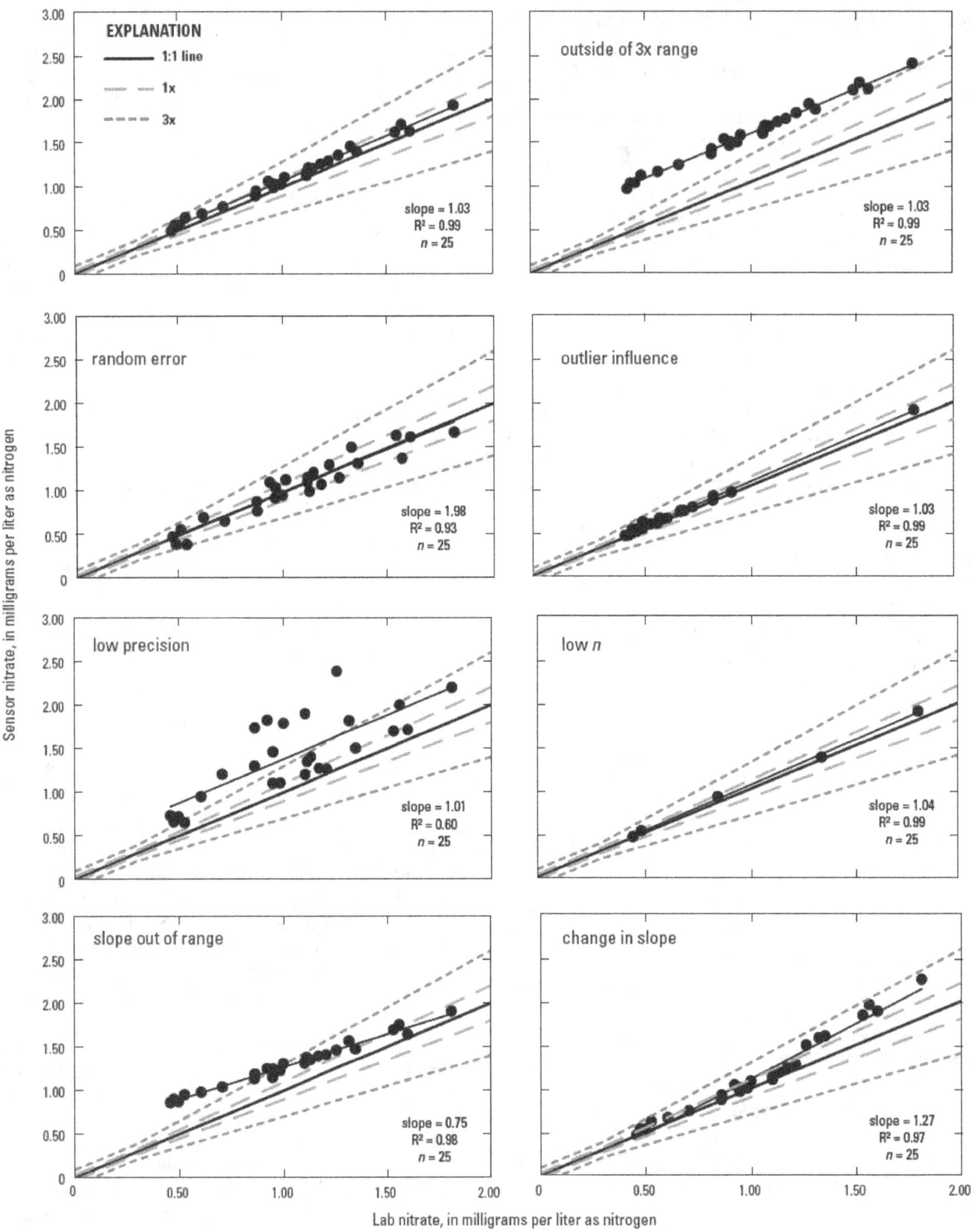

Figure 16. Hypothetical examples of relationships between lab and in situ sensor nitrate concentrations that show correlation (upper left panel) and violations of the assumptions for making a bias correction. The 1 to1 line (solid black), 1x calibration criteria line (green dashed) and 3x calibration criteria line (red dashed) are based on the manufacturer specification for a SUNA sensor.

Table 8. Maximum error for bias corrections of ultraviolet nitrate-sensor data.

[**Abbreviations**: mg/L, milligrams per liter; mm, millimeter; %, percent]

Instrument	Maximum error for bias correction (3x manufacturer accuracy specifications)
Hach Nitratax	±10–15% of reading or ±1.5–3.0 mg/L, whichever is greater.
Satlantic SUNA	±30% of reading or ±0.09–0.18 mg/L, whichever is greater.
S::CAN spectrolyzer	±6% of reading plus 3/optical path length (in mm; mg/L).
TriOS ProPS	±6% of reading or ±0.465 mg/L, whichever is greater.

Reporting Parameter and Method Codes

All concentration data from UV nitrate sensors can be stored in NWIS under the appropriate parameter codes. Several parameter codes have historically been used for nitrate sensor measurements (see *http://nwis.waterdata.usgs.gov/usa/nwis/pmcodes* for a current list of parameter codes), but the future use of parameter code 99133 ("nitrate plus nitrite, water, in situ, milligrams per liter as nitrogen") can provide consistent reporting of continuous in situ UV nitrate sensor concentrations. The use of this parameter code alleviates several unnecessary or ambiguous terms often used to describe the field nitrate measurement. The following are examples:

- The parameter code describes the measurement as "nitrate plus nitrite" because nitrite, which absorbs strongly in the range of 210–220 nm, is not explicitly accounted for in the nitrate calculations by the sensors. If an algorithm is developed or used to remove the contribution of nitrite, the use of parameter code 99137 ("nitrate, water, in situ, milligrams per liter as nitrogen") is appropriate. For practical purposes, the concentration of nitrite is almost always negligible in surface waters and has little effect on reported nitrate concentrations.

- We use the term "in situ" instead of "field" because there are a wide range of alternative uses for "field" in the NWIS database that are not associated with continuous water-quality sensor measurements.

- Reporting as units of mg/L as N, rather than as molar units or reporting concentrations as nitrate is in more common use and facilitates comparisons among studies; therefore, it is used for reporting nitrate concentrations, unless specifically required otherwise by a study.

- The term "filtered" is not necessary to the parameter code for in situ measurements because nitrate is a dissolved constituent. Although "filtered" can indicate a method of field collection (that is, pumping water through a filter prior to measurement and thereby eliminating particle interference), the application of bias corrections for matrix effects makes this process unnecessary.

The use of method codes for UV nitrate sensors that identify the method, sensor manufacturer, the model, and the instrument path length is the most informative. Table 9 includes the parameter codes for many of the UV nitrate sensors currently used by the USGS, but new method codes can be requested for sensors or instrument path lengths not currently on this list. In addition, method codes containing the same descriptive information can also be established for wet chemical nitrate sensors.

Final Data Evaluation and Review

Final data evaluations and review for UV nitrate measurements follow the general guidelines of Wagner and others (2006) for record checking, station analysis, data reporting, and archiving of continuous water-quality data. However, two concepts that have been applied to other types of continuous monitors—accuracy ratings and maximum allowable limits—are further discussed for UV nitrate sensors. Accuracy ratings for continuous data are required by the USGS for data reports and other publications.

Table 9. Method codes for continuous ultraviolet (UV) nitrate measurements used by the U.S. Geological Survey for parameter code 99133.

[**Abbreviation**: mm, millimeter]

Method code (example)	Method description (method, manufacturer, path length)
UV012	UV absorption, Satlantic SUNA, 10-mm path length.
UV013	UV absorption, Satlantic SUNA, 5-mm path length.
UV014	UV absorption, Satlantic ISUS, 10-mm path length.
UV015	UV absorption, HACH Nitratax plus sc, 5-mm path length.
UV016	UV absorption, HACH Nitratax plus sc, 2-mm path length.
UV017	UV absorption, s::can spectrolyzer, 5-mm path length.
UV018	UV absorption, s::can spectrolyzer, 15-mm path length.

For UV nitrate sensors, accuracy ratings for the absolute sum of the corrections (fouling, drift, and bias) based on the manufacturer-stated accuracy specifications can be used (table 10). The use of individual instrument accuracy specifications as criteria for assigning data ratings means that instruments with better accuracy have a greater likelihood of receiving lower ratings. Therefore, it is important to exercise extreme caution when comparing different instruments on the basis of data ratings, and to evaluate whether alternate ways of assigning rating criteria are necessary for a given study or instrument.

The maximum allowable limit—defined as the maximum value for the absolute sum of all data corrections where data are reported, stored in the NWIS database, or both (Wagner and others, 2006)—has important implications for UV nitrate sensors. If the limit is too large, the ability to interpret results is compromised; if the limit is too low, the costs of instrument servicing and field visits can be excessive. Six times the manufacturer-stated (table 2) or user-defined accuracy is an appropriate maximum allowable limit for drift, fouling, and bias corrections. However, users could evaluate the data quality when rating accuracies of fair (within three to four times the sensor accuracy) or poor (within four to six times the sensor accuracy) are reported because the need for such large corrections can be indicative of sensor lamp degradation, wiper failure, significant matrix interferences, or user error. Data that exceed established criteria or are determined to be unacceptable for publication or storage can be flagged in NWIS and archived according to Water Science Center quality-assurance plans.

Table 10. Accuracy ratings based on the absolute sums of the combined fouling, drift, and bias corrections to discrete samples for continuous ultraviolet nitrate measurements.

[The sensor accuracy used can be manufacturer-stated specifications or user-calculated values.]

Accuracy rating	Specification
Excellent	Within sensor accuracy.
Good	±1–3 times sensor accuracy.
Fair	±3–4 times sensor accuracy.
Poor	±4–6 times sensor accuracy.

Summary

The recent commercial availability of in situ optical sensors, together with new techniques for data collection and analysis, provide the opportunity to monitor water quality on the time scales in which the environmental changes. In particular, UV photometers to determine in situ nitrate concentrations by absorbance measurements are sufficiently developed to warrant their broader application; however, generating data that meet high USGS standards requires an investment in and adherence to common methods and protocols for sensor selection, characterization, and operation, as well as for data-quality assurance, control, and management.

The goal of this report is to provide information that helps USGS personnel collect reproducible nitrate concentration data by using UV sensors in ways that are comparable across sites and studies; thus, this report covers a broad range of topics, from choosing the right sensor to maintaining sensors in the field. While many of these topics overlap with guidelines for more traditional chemical and physical water-quality sensors, other topics, such as matrix interferences from suspended particles and colored organic matter, are unique to optical instruments. Individual sensor selection can be determined by the expected range in nitrate and matrix elements (such as DOC and suspended sediments), study specifications, or reporting limits for accuracy and precision, and logistical constraints. Differences among instruments, such as path lengths and wavelengths measured, are critical features that affect data quality and can be considered along with design differences that affect the depth rating, temperature rating, and maintenance.

Field deployment and maintenance protocols ensure that sensors are working properly and provide critical data for interpreting and processing the water-quality record. Although standard protocols defined by Wagner and others (2006) for continuous monitors apply to the field servicing of UV nitrate sensors, several important differences between UV nitrate sensors and many of the traditional continuous water-quality probes such as pH and temperature require additional steps. For example, UV photometers have a sensitive, high-powered lamp that degrades over time and can be compensated for directly by the instrument (for those with a reference channel) or by the user. In addition, the desired output (nitrate concentration) is derived from an algorithm that can be influenced by a variety of dissolved inorganic and organic substances and particles in the sample path. Finally, the new generation of sensors provides opportunities to collect diagnostic information that can improve the ability to evaluate instrument performance and data quality. Existing protocol for data corrections due to fouling and drift of continuous water-quality monitors apply to UV nitrate sensors, but additional corrections for systematic errors due to matrix effects can also be justified.

Acknowledgments

The authors are grateful for USGS funding and support from the National Water Quality Assessment (NAWQA) program, the Office of Water Quality (OWQ), the Office of Surface Water (OSW), the National Stream Quality Accounting Network (NASQAN), the Hydrologic Instrumentation Facility (HIF) and the National Monitoring Network for Coastal Waters and their Tributaries. The authors also gratefully acknowledge the USGS Ad Hoc Optical Sensor Group for technical contributions to this document. Members of the group include Ken Hyer, Pat Rasmussen, Stewart Rounds, Rick Wagner, Yvonne Stoker, George Ritz, Terri Snazzelle, Dan Sullivan, Stan Skrobialowski, and Franceska Wilde. We thank Jeannie Barlow, Joseph Bell, Shawn Fisher, Dick Cartwright, Jack Gibs, Janice Fulford, Stephen Huddleston, and members of the USGS Continuous Water Quality Committee for helpful comments and discussions. We also thank Geoff McIntyre, Justin Irving, Nichole Halsey, Patrick Barnard, and Sven-Erik Krause for help with instrument specifications. Finally, we thank Donna Myers, Bill Wilber, Charlie Crawford, Bob Gilliom, Robert Mason and Pete Murdoch for their leadership and vision for the application of in situ optical sensors by the USGS.

References Cited

American Public Health Association, American Water Works Association, and Water Environment Federation, 1995, Standard methods for the examination of water and wastewater, (19th ed.): American Public Health Association, American Water Works Association, and Water Environment Federation Washington, D.C., p. 4-85–86.

Armstrong F.A.J., 1963, Determination of nitrate in water by ultraviolet spectrophotometry: Analytical Chemistry, v. 35, no. 9, p. 1292–1293.

Bastin R., Weberling, T., and Padilla, F., 1957, Ultraviolet spectrophotometric determination of nitrate: Analytical Chemistry, v. 29, p. 1795–1797.

Bryan N.S. and Loscalzo, J., eds, 2011, Nitrate and nitrite in human health and disease: New York, Springer, Humana Press, 306 p.

Cohen M.J., Kurz, M.J., Heffernan, J.B., Martin, J.B., Douglass, R.L., Foster, C.R., and Thomas, R.G., 2013, Diel phosphorus variation and the stoichiometry of ecosystem metabolism in a large spring-fed river: Ecological Monographs, v. 83, p. 155-176.

Drolc, A. and Vrtovšek, J., 2010, Nitrate and nitrite nitrogen determination in waste water using on-line UV spectrometric methods: Bioresource Technology, v. 101, p. 4228–4233.

Finch, M.S., Hydes, D.J., Clayson, C.H., Weigl, B.H., Dakin, J.P., and William, P.G., 1998, A low power ultra violet spectrophotometer for measurement of nitrate in seawater: Introduction, calibration and initial sea trials: Analytica Chimica Acta, v. 377, no. 2–3, p. 167–177.

Heffernan J.B. and Cohen, M.J., 2010, Direct and indirect coupling of primary production and diel nitrate dynamics in a subtropical spring-fed river: Limnology and Oceanography, v. 55, no. 2, p. 677–688.

Hu, C., Muller-Karger, F.E., Zepp, R.G., 2002, Absorbance, absorption coefficient, and apparent quantum yield: A comment on common ambiguity in the use of these optical concepts: Limnology and Oceanography, v. 47, no. 4, p. 1262–1267.

Johnson K.S., 2010, Nitrate supply from deep to near-surface waters of the North Pacific subtropical gyre: Nature, v. 465, p. 1062–1065.

Johnson K.S. and Coletti, L.J., 2002, In situ ultraviolet spectrophotometry for high resolution and long-term monitoring of nitrate, bromide and bisulfide in the ocean: Deep Sea Research, v. 49, p.1291–1305.

Johnson K.S., Coletti, L.J., and Chavez , F.P., 2006, Diel nitrate cycles observed with in situ sensors predict monthly and annual new production: Deep Sea Research, v. 53, p. 561–573.

Lane S.L. and Fay, R.G., 1997, Safety in field activities: U.S. Geological Survey Techniques of Water-Resources Investigations, book 9, chap. A9, October 1997, accessed October 1, 2012, at *http://pubs.water.usgs.gov/twri9A9/*.

Lane, S.L., Flanagan, S., and Wilde, F.D., 2003, Selection of equipment for water sampling (ver. 2.0): U.S. Geological Survey Techniques of Water-Resources Investigations, book 9, chap. A2, March 2003, accessed August 6, 2012, at *http://pubs.water.usgs.gov/twri9A2/*.

Langergraber, G., Gupta, J.K., Pressl, A., Hofstaedter, F., Lettl, W., Weingartner, A., and Fleischmann, N., 2004, On-line monitoring for control of a pilot-scale sequencing batch reactor using a submersible UV/VIS spectrometer: Water Science and Technology, v. 50, no. 10, p. 73–80.

Patton C.J. and Kryskalla, J.R., 2011, Colorimetric determination of nitrate plus nitrite in water by enzymatic reduction, automated discrete analyzer methods: U.S. Geological Survey Techniques and Methods, book 5, chap. B8, 34 p.

Pellerin B.A., Bergamaschi, B.A., and Horsburgh, J.S., 2012, In situ optical water-quality sensor networks—Workshop summary report: U.S. Geological Survey Open-File Report 2012–1044, 13 p.

Pellerin B.A., Downing, B.D., Kendall, C., Dahlgren, R.A., Kraus, T.E.C., Spencer, R.G., and Bergamaschi, B.A., 2009, Assessing the sources and magnitude of diurnal nitrate variability in the San Joaquin River (California) with an in situ optical nitrate sensor and dual nitrate isotopes: Freshwater Biology, v. 54, p. 376–387.

Pellerin, B.A., Saraceno, J., Shanley, J.B., Sebestyen, S.D., Aiken, G.R., Wollhim, W.M., and Bergamaschi, B.A., 2011, Taking the pulse of snowmelt: in situ sensors reveal seasonal, event and diurnal patterns of nitrate and dissolved organic matter variability in an upland forest stream: Biogeochemistry, DOI 10.1077/s10533-011-9589-8.

Reiger, L., Langergraber, G., Kaelin, D., Siegrist, H., and Vanrolleghem, P.A., 2008, Long-term evaluation of a spectral sensor for nitrite and nitrate: Water Science and Technology, v. 57, no. 10, p. 1563–1569.

Roesler C., 1998, Theoretical and experimental approaches to improve the accuracy of particulate absorption coefficients derived from the quantitative filter technique: Limnology and Oceanography, v. 43, no. 7, p. 1649–1660.

Sakamoto, C.M., Johnson, K.S., and Coletti, L.J., 2009, Improved algorithm for the computation of nitrate concentrations in seawater using an *in situ* ultraviolet spectrophotometer: Limnology and Oceanography: Methods, v. 7, p. 132–143.

Sanford, R.C., Exenberger, A., and Worsfold, P.J., 2007, Nitrogen cycling in natural waters using in situ, reagentless UV spectrophotometry with simultaneous determination of nitrate and nitrite: Environmental Science and Technology, v. 41, p. 8420–8425.

Sathyendranath, S., Lazzara, L., and Prieur, L., 1987, Variations in the spectral values of specific absorption of phytoplankton: Limnology and Oceanography, v. 32, no. 403–415.

Skoog, D.A., 1985, Principles of instrumental analysis (3d ed.): Philadelphia, Saunders College Pub., 879 p.

Townsend, A.R., Howarth, R.W., Bazzaz, F.A., Booth, M.S., Cleveland, C.C., Collinge, S.K., Dobson, A.P., Epstein, P.R., Holland, E.A., Keeney, D.R., Mallin, M.A., Rogers, C.A., Wayne, P., and Wolfe, A.H., 2003, Human health effects of a changing global nitrogen cycle: Frontiers in Ecology and Environment, v. 1, no. 5, p. 240–246.

U.S. Geological Survey, variously dated, National field manual for the collection of water-quality data: U.S. Geological Survey Techniques of Water-Resources Investigations, book 9, chaps. A1–A9, available online at *http://pubs.water.usgs.gov/twri9A.*

U.S. Geological Survey, 2006, Collection of water samples (ver. 2.0): U.S. Geological Survey Techniques of Water-Resources Investigations, book 9, chap. A4, September 2006, accessed August 6, 2012, at *http://pubs.water.usgs.gov/twri9A4/.*

Vitousek, P.M., Aber, J.D., Howarth, R.W., Likens, G.E., Matson, P.A., Schindler, D.W., Schlesinger, W.H., and Tilman, D.G., 1997, Human alternation of the global nitrogen cycle: Sources and consequences: Ecological Applications, v. 7, p. 737–750.

Wagner R.J., Boulger Jr., R.W., Oblinger, C.J., and Smith, B.A., 2006, Guidelines and standard procedures for continuous water-quality monitors—Station operation, record computation, and data reporting: U.S. Geological Survey Techniques and Methods 1–D3, 51 p. + 8 attachments; accessed August 6, 2012, at *http://pubs.water.usgs.gov/tm1d3.*

Wilde, F.D., ed., 2004, Cleaning of equipment for water sampling (ver. 2.0): U.S. Geological Survey Techniques of Water-Resources Investigations, book 9, chap. A3, April 2004, accessed August 6, 2012, at *http://pubs.water.usgs.gov/twri9A3/.*

Zielinski, O., Voß, D., Saworski, B., Fiedler, B., and Körtzinger, A., 2011, Computation of nitrate concentrations in turbid coastal waters using an in situ ultraviolet spectrophotometer: Journal of Sea Research, v. 65, p. 456–460.

Appendix 1. Calculating Data-Quality Specifications for UV Nitrate Sensors

Performance Qualifications

The information presented here is intended to help users evaluate ultraviolet (UV) nitrate sensor accuracy, precision, and linearity, as detailed in the "Instrument Performance Qualifications" section of this report (Operation Inspection, d and e). Although not detailed in the "Instrument Performance Qualifications" section, the approach for determining or verifying the method detection limit for UV nitrate sensors is included here. (Calculations for instrument bias are presented in the report itself). Users can carefully review the manufacturer manual on instrument operation and stated instrument specifications before doing the evaluations.

The performance of UV nitrate sensors can be verified by using inorganic-free blank water (IBW), nitrate standards, and nitrate spike solutions in natural matrix waters under controlled laboratory conditions. The U.S. Geological Survey (USGS) National Field Supply Service (NFSS) sells IBW as stock number Q378FLD, and the USGS National Water Quality Lab (NWQL) provides results of the acceptance testing for each lot sold *http://wwwnwql.cr.usgs.gov/qas.shtml?ibw*). If using in-house blank water systems, electrical resistivity greater than 18 mega-ohms per centimeter at 25 degrees Celsius (°C, electrical conductivity less than 0.056 µS/cm at 25°C) is specified and users are to verify that the water is essentially free of inorganic constituents.

Calculating Instrument Accuracy

Accuracy is the degree of agreement between the measured nitrate value and its known concentration. The accuracy of an instrument is usually assessed by comparison of measured values to known values in standard solutions. Accuracy specifications are reported by manufacturers (table 2), but are normally determined by using nitrate standards in IBW free of nitrate or potential interferences such as bromide or color. In the "Instrument Performance Qualifications," we described assessing the instrument accuracy in both nitrate standard solutions and nitrate spikes to environmental samples from the intended deployment site prior to deployment.

Nitrate concentrations can be measured in standard solutions, or spike solutions in natural matrix water, across the range of concentrations (typically three to five samples) typical for the intended deployment site. The spike concentrations can be one to five times the background nitrate concentration, or equal or below any regulatory limit or action level, whichever is greater. Users can collect sufficient environmental matrix samples to determine instrumental performance over the range interferences (dissolved organic matter, turbidity, or bromide) expected to be encountered at the field site, or use standard reference materials (see "Instrument Performance Qualifications") to generate equivalent conditions in the lab.

UV nitrate sensor accuracy can be calculated in the laboratory through the following steps:

1. Measure the background nitrate concentration in the IBW or an unspiked matrix sample (C_u) by using the sensor as described in the user manual.

2. For accuracy calculations that use nitrate standards in IBW, either purchase or make a stock standard for use in instrument testing. Nitrate standards are available from a range of vendors and at a range of concentrations. Follow directions for sample handling, storage, and shelf life.

3. For nitrate spikes in matrix waters, the theoretical (C_c) concentration from spiking can be calculated from the following equation:

$$C_c = \frac{C_{std} * V_{std}}{V_{sample}} \qquad (1\text{-}1)$$

where

C_c is the calculated or target concentration from spiking the sample (typically in milligrams per L, mg/L, as nitrogen, N, for nitrate sensors),

C_{std} is the concentration of the nitrate spike standard (in mg/L as N),

V_{std} is the volume added of the nitrate spike standard (in milliliters, mL), and

V_{sample} is the total volume of the spiked sample (L).

The equation can be rearranged to solve for the volume of known nitrate standard to be added for a matrix spike at a desired concentration:

$$V_{std} = \frac{C_c * V_{sample}}{C_{std}}$$

4. Calculate The accuracy of the measurement of the spike solution is calculated on the basis of the percentage or absolute recovery of added nitrate by using the following equations:

$$Recovery(\%) = R\% = \frac{C_s \quad C_u}{C_c} * 100 \text{ percent} \qquad (1\text{-}2)$$

$$Recovery(conc) = (C_s \quad C_u) \quad C_c \qquad (1\text{-}3)$$

where

C_s is the measured concentration of spiked sample or nitrate standard,

C_u is the measured concentration of unspiked sample or IBW, and

C_c is the calculated concentration of the nitrate standard or matrix spike.

The nitrate concentration in IBW, C_u, is expected to be zero, and accuracy (that is, recovery percentage, R%) is expressed as the measured value (C_s) divided by the known value (C_c) multiplied by 100 percent. Note, however, that the C_u in IBW can be a negative value with some instruments. If this is the case, users can use the negative value as long as it falls within the calibration criteria (see table 5) for the previous calculations, but users can also re-evaluate the blank values with IBW for evidence of sensor drift or damage.

The following is an example of calculating instrument accuracy with a single value. A matrix sample collected in the field is measured by using the UV nitrate sensor in the laboratory and determined to have a background nitrate concentration of 0.55 mg/L as N. To obtain a desired spike concentration three times the background concentration (that is, 1.65 mg/L as N) by using a nitrate standard solution with a certified or known concentration of 1,000 mg/L as N, the volume of standard solution needed (V_{std}) to make a 200 mL spike sample is calculated as follows:

$$V_{std} = \frac{C_c * V_{sample}}{C_{std}} = \frac{1.65 \text{ mg/L as N} * 200 \text{ mL}}{1{,}000 \text{ mg/L as N}} = 0.33 \text{ mL}$$

After adding 0.33 mL of nitrate standard to a volumetric flask and bringing the final volume up to 200 mL with matrix water, the spiked sample is measured with the sensor (C_s) as 2.40 mg/L as N. The percent and absolute recoveries are calculated as follows:

$$Recovery(\%) = \frac{2.40 - 0.55}{1.65} * 100\, percent = 112\%$$

$$Recovery(conc) = (2.40 \quad 0.55) \quad 1.65 = 0.20 \text{ mg/L as N}$$

On the basis of these values, the manufacturer-stated accuracy can be verified, or user-specified instrument accuracy can be calculated. Finding a systematic bias in natural waters—typically an overestimate in nitrate concentrations—is likely to be indicative of a matrix interference. A systematic bias in nitrate standards can be indicative of the need for a baseline correction or a sensor performance problem.

Instrument Linearity Check

The UV absorption method for nitrate is a direct spectrophotometric measurement, and UV nitrate sensor concentrations, therefore, are expected to be linearly correlated with laboratory nitrate measurements in the absence of interferences. Linearity checks confirm that the instrument response is linear for nitrate standards and spike solutions. during accuracy checks, if nitrate concentrations span a range typical for the intended deployment site, the concentration values can be used to assess linearity of the sensor response. If the sensor accuracy is within specifications (as stated by the manufacturer or calculated by the user) at 3–5 different nitrate concentrations across the range, linearity typically can be assumed. However, use of least squares linear regression could be a more appropriate measure of linearity. The linear regression equation takes the following form:

$$Y = mx + b \qquad (1\text{-}4)$$

where

m is the slope of the line, and
b is the y-intercept.

The slope (m) of a regression between sensor measurements and laboratory nitrate standards is expected to be one, in the absence of interferences, and to have an intercept (b) near zero. Deviation of the intercept from zero could be indicative of a systematic offset or bias in the relationship and can be further evaluated to assess for instrument drift (possibly requiring a baseline correction) or matrix effects. The coefficient of determination (R^2) also approaches one when there is a strong linear relationship between sensor measurements and nitrate standard concentrations, and R^2 values less than 0.95 can be further investigated for outliers or non-linearity in instrument output.

The following is an example of a linearity check using standard solutions. A series of five nitrate standards—0, 0.1 0.5, 1.0, and 2.0 mg/L as N—in IBW were measured with a nitrate sensor in the laboratory. The data (fig. 1-1) indicate linearity with a slope and R^2 value near one and an intercept near zero.

Precision

Although it is not a component of the "Instrument Performance Qualification" section, users can verify or determine instrument precision by using repeated measurements of nitrate standard and in spike matrix samples during accuracy assessments. (Instrument manufacturers provide an estimate of precision, but this can be used for validation or for an assessment under typical matrix conditions). A statistical measure of the precision for a series of repeated measurements is the standard deviation (S):

$$S = \sqrt{\frac{\sum_{i=1}^{N}(X_i \quad \bar{X})^2}{n \quad 1}} \qquad (1\text{-}5)$$

where

S is the standard deviation (in the same units as the measurements),
n is the number of measurements,
X_i is each individual measurement, and
$\bar{X}$ is the mean of all measurements.

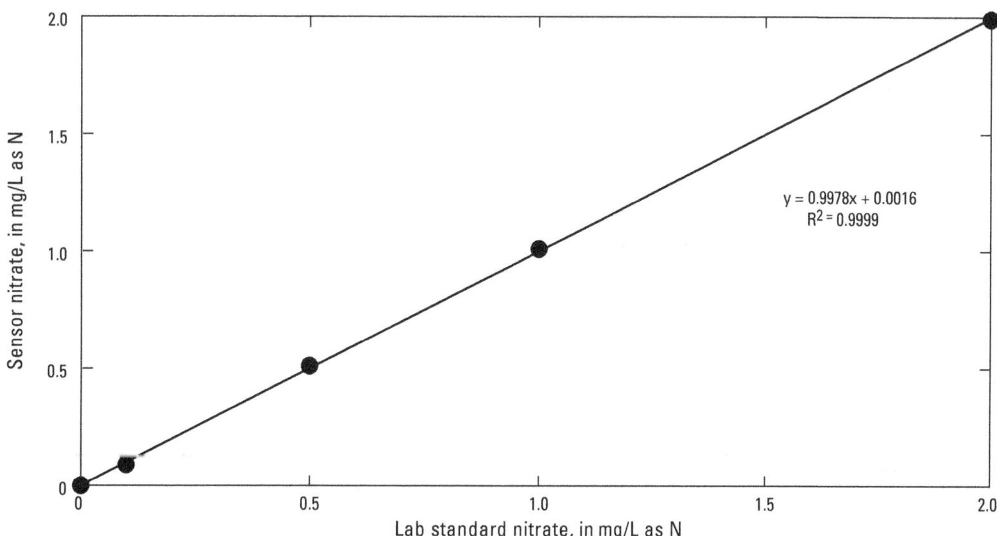

Figure 1-1. Example linearity check using standard nitrate solutions (milligrams per liter as nitrogen, mg/L as N) measured with laboratory instruments and nitrate sensors.

The following is an example of calculating precision during matrix spikes. An average nitrate concentration of 2.40 mg/L as N was recorded during a lab matrix spike during a 30-second sampling period with a sampling rate of 2 seconds (that is, 15 measurements taken, table 1-1). The sample has a standard deviation of ±0.06 mg/L as N, which can also be expressed as a relative standard deviation in percent (%):

$$RSD(\%) = \frac{S}{\overline{X}} * 100 = \frac{0.06}{2.40} * 100\, percent = 2.5\%$$

This example is the calculation of the sample precision because the repeated measurements were made in spiked matrix waters. Manufacturer stated precision is typically calculated in IBW or nitrate standards in the absence of interferences and, therefore, can be considered analytic precision rather than sample precision. In addition, some manufacturers could calculate this differently. Check user manuals or contact manufacturers for clarification on how the stated precision is calculated, if needed.

Table 1-1. Example data for calculating instrument precision during matrix spikes.

[**Abbreviations**: mg/L as N, milligram per liter as nitrogen; $\overline{X}$, mean of all measured sensor concentrations; X_i, measured sensor concentration, in milligrams per liter as nitrogen]

Sensor measured concentration, X_i (mg/L as N)	X_i-mean, $\overline{X}$ (mg/L as N)
2.40	0.00
2.44	0.04
2.30	–0.10
2.40	0.00
2.38	–0.02
2.50	0.10
2.38	–0.02
2.54	0.14
2.30	–0.10
2.35	–0.05
2.45	0.05
2.40	0.00
2.42	0.02
2.37	–0.03
2.40	0.00
2.40	mean (mg/L as N)
Standard deviation (mg/L as N)	0.06

Method Detection Limit

The method detection limit (MDL) is the minimum concentration that can be measured and reported as greater than zero at the 99 percent confidence level (U.S. Environmental Protection Agency, 1997). The intent of the MDL is to limit incorrect reporting of the presence of a compound at low concentrations (a "false positive") to less-than or equal to one percent. The MDL is typically determined through the analysis of variability of multiple samples by using either low-concentration nitrate standards in IBW or blind blanks (Childress and others, 1999). However, the MDL can be determined for UV nitrate sensors by using either repeated sensor measurements in IBW or nitrate standards at one to five times the expected MDL (typically based on manufacturer specifications) in blank water. A minimum of seven replicate measurements (*n*) are needed according to the US EPA-prescribed method for MDL determination, which is calculated as follows:

$$MDL = T_{(n\ 1,1\ \alpha=0.99)} * S \quad (1\text{-}6)$$

where

$T_{(n-1,1-\alpha=0.99)}$ is the student's t value (one-tailed) at the 99 percent confidence level with *n*-1 degrees of freedom, and

S is the standard deviation (calculated as described for precision).

The following is an example of calculating the MDL. A nitrate standard at 0.1 mg/L as N (one to five times the reported MDL for most UV nitrate sensors) was measured over a 30 second period during laboratory testing with a sampling rate of 2 seconds (that is, 15 measurements collected). The standard deviation is calculated from the data in table 1-2 on basis of equation 1–5 for calculating precision:

Table 1-2. Example data for calculating instrument detection limits.

[**Abbreviations**: mg/L as N, milligram per liter as nitrogen; $\overline{X}$, mean of all measured sensor concentrations; X_i, measured sensor concentration, in milligrams per liter as nitrogen]

Sensor measured concentration, X_i (mg/L as N)	X_i-mean, $\overline{X}$ (mg/L as N)
0.10	0.00
0.11	0.01
0.08	–0.02
0.09	–0.01
0.09	–0.01
0.11	0.01
0.09	–0.01
0.10	0.00
0.12	0.02
0.08	–0.02
0.09	–0.01
0.08	–0.02
0.10	0.00
0.09	–0.01
0.11	0.01
0.10	mean (mg/L as N)
Standard deviation (mg/L as N)	0.01

The student t-value is determined for the 99 percent confidence interval and 14 (that is, n-1) degrees of freedom by using a statistical distribution table:

$$T_{(15\ 1,1\ \alpha=0.99)} = 2.624$$

Consequently, the method detection limit (MDL) is calculated as follows:

$$MDL = 2.624 * 0.01 = 0.03 \text{ mg/L as N}$$

Because the MDL is typically measured in solutions without any matrix interferences that are likely to affect the true detection limits in natural waters, users could wish to determine the MDL in a more complex matrix typical of the intended deployment site to better understand the instrument performance under field conditions. The calculated MDL is reported to the appropriate number of significant digits on the basis of the measurement resolution of the sensor. For example, a calculated MDL of 0.14 mg/L as N is reported as 0.1 mg/L as N for a sensor that only reports nitrate concentration to one decimal.

References Cited

Childress, C.J. Oblinger; Foreman, W.T.; Connor B.F.; and Maloney, T.J., 1999, New reporting procedures based on long-term method detection levels and some considerations for interpretations of water-quality data provided by the U.S. Geological Survey National Water Quality Laboratory: U.S. Geological Survey Open-File Report, 99–193, 19 p.

U.S. Environmental Protection Agency, 1997, Guidelines establishing test procedures for the analysis of pollutants (App. B, Part 136, Definition and procedures for the determination of the method detection limit): U.S. Code of Federal Regulations, Title 40, revised July 1, 1997, p. 265–267.

www.ingramcontent.com/pod-product-compliance
Lightning Source LLC
Chambersburg PA
CBHW081403170526
45166CB00010B/3188
9781500220228